176
Advances in Polymer Science

Editorial Board:
A. Abe · A.-C. Albertsson · R. Duncan · K. Dušek · W. H. de Jeu
J. F. Joanny · H.-H. Kausch · S. Kobayashi · K.-S. Lee · L. Leibler
T. E. Long · I. Manners · M. Möller · O. Nuyken · E. M. Terentjev
B. Voit · G. Wegner

Advances in Polymer Science

Recently Published and Forthcoming Volumes

**Poly(acrylene Ethynylenes) –
From Synthesis to Application**
Volume Editor: Weder, C.
Vol. 177, 2005

Metathesis Polymerization
Volume Editor: Buchmeiser, M.
Vol. 176, 2005

Polymer Particles
Volume Editor: Okubo, M.
Vol. 175, 2005

Neutron Spin Echo in Polymer Systems
Volume Editors: Richter, D.,
Monkenbusch, M., Arbe, A., Colmenero, J.
Vol. 174, 2005

**Advanced Computer Simulation
Approaches for Soft Matter Sciences I**
Volume Editors: Holm, C., Kremer, K.
Vol. 173, 2004

Microlithography · Molecular Imprinting
Volume Editor: Kobayashi, T.
Vol. 172, 2005

Polymer Synthesis
Vol. 171, 2004

**NMR · Coordination Polymerization ·
Photopolymerization**
Vol. 170, 2004

Long-Term Properties of Polyolefins
Volume Editor: Albertsson, A.-C.
Vol. 169, 2004

Polymers and Light
Volume Editor: Lippert, T. K.
Vol. 168, 2004

New Synthetic Methods
Vol. 167, 2004

**Polyelectrolytes with Defined
Molecular Architecture II**
Volume Editor: Schmidt, M.
Vol. 166, 2004

**Polyelectrolytes with Defined
Molecular Architecture I**
Volume Editors: Schmidt, M.
Vol. 165, 2004

**Filler-Reinforeced Elastomers ·
Scanning Force Microscopy**
Vol. 164, 2003

**Liquid Chromatography ·
FTIR Microspectroscopy · Microwave
Assisted Synthesis**
Vol. 163, 2003

**Radiation Effects on Polymers
for Biological Use**
Volume Editor: Kausch, H.
Vol. 162, 2003

**Polymers for Photonics
Applications II**
Nonlinear Optical, Photorefractive and
Two-Photon Absorption Polymers
Volume Editor: Lee, K.-S.
Vol. 161, 2003

Filled Elastomers · Drug Delivery Systems
Vol. 160, 2002

**Statistical, Gradient, Block
and Graft Copolymers by Controlled/
Living Radical Polymerizations**
Authors: Davis, K. A., Matyjaszewski, K.
Vol. 159, 2002

**Polymers for Photonics
Applications I**
Nonlinear Optical and
Electroluminescence Polymers
Volume Editor: Lee, K.-S.
Vol. 158, 2002

Degradable Aliphatic Polyesters
Volume Editor: Albertsson, A.-C.
Vol. 157, 2001

**Molecular Simulation · Fracture ·
Gel Theory**
Vol. 156, 2001

Metathesis Polymerization

Volume Editor: Michael R. Buchmeiser

With contributions by
T. W. Baughman · M. R. Buchmeiser · G. Fuchs · S. Riegler
C. Slugovc · F. Stelzer · G. Trimmel · K. B. Wagener

Springer

The series presents critical reviews of the present and future trends in polymer and biopolymer science including chemistry, physical chemistry, physics and material science. It is addressed to all scientists at universities and in industry who wish to keep abreast of advances in the topics covered.

As a rule, contributions are specially commissioned. The editors and publishers will, however, always be pleased to receive suggestions and supplementary information. Papers are accepted for "Advances in Polymer Science" in English.

In references Advances in Polymer Science is abbreviated Adv Polym Sci and is cited as a journal.

The electronic content of APS may be found springerlink.com

Library of Congress Control Number: 2004112377

ISSN 0065-3195
ISBN 3-540-23358-X **Springer Berlin Heidelberg New York**
DOI 10.1007/b101315

This work is subject to copyright. All rights are reserved, whether the whole or part of the material is concerned, specifically the rights of translation, re-printing, re-use of illustrations, recitation, broadcasting, reproduction on microfilms or in any other ways, and storage in data banks. Duplication of this publication or parts thereof is only permitted under the provisions of the German Copyright Law of September 9, 1965, in its current version, and permission for use must always be obtained from Springer-Verlag. Violations are liable to prosecution under the German Copyright Law.

Springer is a part of Springer Science+Business Media
springeronline.com
© Springer-Verlag Berlin Heidelberg 2005
Printed in The Netherlands

The use of registered names, trademarks, etc. in this publication does not imply, even in the absence of a specific statement, that such names are exempt from the relevant protective laws and regulations and therefore free for general use.

Cover design: KünkelLopka GmbH, Heidelberg/design & production GmbH, Heidelberg
Typesetting: Fotosatz-Service Köhler GmbH, Würzburg

Printed on acid-free paper 02/3141 xv – 5 4 3 2 1 0

Volume Editor

Prof. Dr. Michael R. Buchmeiser
Institute for Analytical Chemistry and Radiochemistry
University of Innsbruck
Innrain 52a
6020 Innsbruck, Austria
michael.r.buchmeiser@uibk.ac.at

Editorial Board

Prof. Akihiro Abe
Department of Industrial Chemistry
Tokyo Institute of Polytechnics
1583 Iiyama, Atsugi-shi 243-02, Japan
aabe@chem.t-kougei.ac.jp

Prof. A.-C. Albertsson
Department of Polymer Technology
The Royal Institute of Technology
S-10044 Stockholm, Sweden
aila@polymer.kth.se

Prof. Ruth Duncan
Welsh School of Pharmacy
Cardiff University
Redwood Building
King Edward VII Avenue
Cardiff CF 10 3XF, United Kingdom
duncan@cf.ac.uk

Prof. Karel Dušek
Institute of Macromolecular Chemistry,
Czech Academy of Sciences of the
Czech Republic
Heyrovský Sq. 2
16206 Prague 6, Czech Republic
dusek@imc.cas.cz

Prof. Dr. W. H. de Jeu
FOM-Institute AMOLF
Kruislaan 407
1098 SJ Amsterdam, The Netherlands
dejeu@amolf.nl
and

Dutch Polymer Institute
Eindhoven University of Technology
PO Box 513
5600 MB Eindhoven, The Netherlands

Prof. Jean-François Joanny
Institute Charles Sadron
6, rue Boussingault
F-67083 Strasbourg Cedex, France
joanny@europe.u-strasbg.fr

Prof. Hans-Henning Kausch
c/o IGC I, Lab. of Polyelectrolytes
and Biomacromolecules
EPFL-Ecublens
CH-1015 Lausanne, Switzerland
kausch.cully@bluewin.ch

Prof. S. Kobayashi
Department of Materials Chemistry
Graduate School of Engineering
Kyoto University
Kyoto 615-8510, Japan
kobayasi@mat.polym.kyoto-u.ac.jp

Prof. Kwang-Sup Lee
Department of Polymer Science &
Engineering
Hannam University
133 Ojung-Dong
Daejeon 306-791, Korea
kslee@mail.hannam.ac.kr

Prof. L. Leibler
Matière Molle et Chimie
Ecole Supèrieure de Physique
et Chimie Industrielles (ESPCI)
10 rue Vauquelin
75231 Paris Cedex 05, France
ludwik.leibler@espci.fr

Prof. Timothy E. Long
Department of Chemistry
and Research Institute
Virginia Tech
2110 Hahn Hall (0344)
Blacksburg, VA 24061, USA
telong@vt.edu

Prof. Ian Manners
Department of Chemistry
University of Toronto
80 St. George St.
M5S 3H6 Ontario, Canada
imanners@chem.utoronto.ca

Prof. Dr. Martin Möller
Deutsches Wollforschungsinstitut
an der RWTH Aachen e.V.
Pauwelsstraße 8
52056 Aachen, Germany
moeller@dwi.rwth-aachen.de

Prof. Oskar Nuyken
Lehrstuhl für Makromolekulare Stoffe
TU München
Lichtenbergstr. 4
85747 Garching, Germany
oskar.nuyken@ch.tum.de

Dr. E. M. Terentjev
Cavendish Laboratory
Madingley Road
Cambridge CB 3 OHE
United Kingdom
emt1000@cam.ac.uk

Prof. Brigitte Voit
Institut für Polymerforschung Dresden
Hohe Straße 6
01069 Dresden, Germany
voit@ipfdd.de

Prof. Gerhard Wegner
Max-Planck-Institut
für Polymerforschung
Ackermannweg 10
Postfach 3148
55128 Mainz, Germany
wegner@mpip-mainz.mpg.de

Advances in Polymer Science
Also Available Electronically

For all customers who have a standing order to Advances in Polymer Science, we offer the electronic version via SpringerLink free of charge. Please contact your librarian who can receive a password for free access to the full articles by registering at:

springerlink.com

If you do not have a subscription, you can still view the tables of contents of the volumes and the abstract of each article by going to the SpringerLink Homepage, clicking on "Browse by Online Libraries", then "Chemical Sciences", and finally choose Advances in Polymer Science.

You will find information about the

– Editorial Board
– Aims and Scope
– Instructions for Authors
– Sample Contribution

at springeronline.com using the search function.

Preface

Clearly illustrated and demonstrated by the entire series of *Advances in Polymer Science*, the area of polymer science is a rapidly developing and growing field, strongly influencing other areas of chemistry. Among other polymerization techniques, those based on metathesis polymerization have experienced significant progress. With a rapidly developing armory of initiators on hand, one is now capable of polymerizing various types of functional monomers by metathesis-based techniques. Thus, ring-opening metathesis polymerization (ROMP) uses strained monomers such as (substituted) norborn-2-enes, norbornadienes, benzbarrelenes, etc. Acyclic diene metathesis (ADMET) polymerization utilizes functional α,ω-dienes. And finally, an almost unlimited number of 1-alkynes as well as 1,6-heptadiynes may be polymerized via 1-alkyne or cyclopolymerization to yield highly conjugated materials. The latest developments in all these areas of metathesis-based polymerizations are summarized in this book. It is designed to attract equally students and advanced scientists working in the areas of polymer science, physical, and organometallic chemistry by providing both extensive background information and up-to-date interdisciplinary knowledge. Special consideration has been given to the literature sections in order to facilitate further reading.

Any edited book strongly depends on the quality of every individual contribution. It was both my privilege and honor to win over such well-known authors. I wish to thank all of them for the time they spent on writing the corresponding chapters and for the unprecedented timely delivery of their contributions. Both their professional attitude and the quality of their manuscripts have made my job as an editor an easy one. With their contributions, I am convinced that we now have a book that represents the state of the art and is comprehensive summary of the status quo in the selected research areas.

What remains to be done is to thank all those who have provided professional help, i.e. *Springer* and in particular *Ms. Ulrike Kreusel*, for her support, encouraging e-mails and patience.

Innsbruck, Fall 2004   Michael R. Buchmeiser

Contents

Recent Advances in ADMET Polymerization
T. W. Baughman · K. B. Wagener 1

Liquid Crystalline Polymers by Metathesis Polymerization
G. Trimmel · S. Riegler · G. Fuchs · C. Slugovc · F. Stelzer 43

Regioselective Polymerization of 1-Alkynes and Stereoselective Cyclopolymerization of α,ω-Heptadiynes
M. R. Buchmeiser 89

Author Index Volumes 101–176 121

Subject Index 141

Recent Advances in ADMET Polymerization

Travis W. Baughman · Kenneth B. Wagener (✉)

George and Josephine Butler Polymer Laboratory, Department of Chemistry, University of Florida, Gainesville, FL32605, USA
wagener@chem.ufl.edu

1	**Introduction**	2
1.1	History of Olefin Metathesis	3
1.2	History of ADMET	5
2	**Functionalized Polyethylene via ADMET: Model Copolymers of Ethylene and Vinyl Monomers**	6
2.1	Ethylene-Propylene Copolymers	7
2.2	Ethylene-Vinyl Acetate Copolymers	10
2.3	Other Ethylene Copolymers	12
2.3.1	Ethylene-Styrene Copolymers	12
2.3.2	Ethylene-Acrylate Copolymers	12
2.3.3	Ethylene-Vinyl Chloride Copolymers	13
3	**Block and Graft Copolymers via ADMET**	15
3.1	Grafted Polyethylenes	15
3.2	Polyethylene-*g*-Poly(Ethylene Glycol)	15
3.3	Polyethylene-*g*-Polystyrene	17
3.4	Block Copolymers via ADMET Polyoctenamer Telechelics	18
3.5	Alternating Copolymers	22
4	**Polymeric Materials via ADMET**	23
4.1	Phosphazene Polymers	23
4.2	Poly(*p*-phenylene vinylene) Oligomers	25
5	**Chiral Polymers via ADMET**	27
5.1	Amino Acid-Containing Polymers	27
5.2	D-*chiro*-Inositol-Based ADMET Polymers	30
6	**Silicon-Containing Polymers**	32
6.1	PDMS-*b*-Polyoctenamer-*b*-PDMS	33
6.2	Carbosilane Polymers via ADMET	34
6.3	Latent Reactive Carbosilane Polymers	35
7	**Conclusion**	37
	References	37

Abstract Acyclic diene metathesis (ADMET) polymerization techniques and methodologies developed over the past five years are reviewed. Through constant catalyst development and further understanding of catalytic activity and side reactions, metathesis has solved a number of synthetic problems through the mild carbon-carbon bond-forming reaction. Polymerization of functionalized α,ω-dienes has afforded strictly linear copolymers of ethylene and various polar monomers that are unattainable through chain polymerization methodology. Telechelic preparation via ADMET allows the synthesis of reactive polymers as starting points for block and segmented copolymers. Application of ADMET to materials synthesis has yielded novel amino acid and peptide polymers as well as silicon-based materials.

Keywords ADMET · Metathesis · Polymerization · Polyethylene · Copolymers

List of Abbreviations

ADMET	Acyclic diene metathesis
ATRP	Atom transfer radical polymerization
CPE	Chlorinated polyethylene
DSC	Differential scanning calorimetry
EEA	Ethylene-ethylacrylate
EMA	Ethylene-methylacrylate
EVA	Ethylene-vinyl acetate
ES	Ethylene-styrene
GPC	Gel permeation chromatography
LALLS	Low-angle laser light scattering
MALDI-TOF	Matrix assisted laser desorption ionization-time of flight
MEM	methoxyethoxymethyl
OLED	Organic light emitting diode
PDMS	Poly(dimethylsiloxane)
PEG	Poly(ethylene glycol)
PPV	Poly(p-phenylene vinylene)
PS	Poly(styrene)
RCM	Ring-closing metathesis
ROMP	Ring-opening metathesis polymerization
TFA	Trifluoroacetic acid
THF	Tetrahydrofuran
VPO	Vapor pressure osmometry

1
Introduction

Olefin metathesis has quickly become one of the most widely used methods for mild carbon-carbon bond formation in organic synthesis [1, 2]. With the development of highly active, functional group-tolerant catalysts, like Grubbs' second generation catalyst ([Ru]*), metathesis has been successfully applied across many areas of research, and some reviews already exist that deal with metathesis catalysis and applications [1–5]. This review focuses on recent developments in acyclic diene metathesis (ADMET) polymerization chemistry and methodology that have been published over the past five years, starting with a short discussion on the history of olefin metathesis and ADMET polymerization.

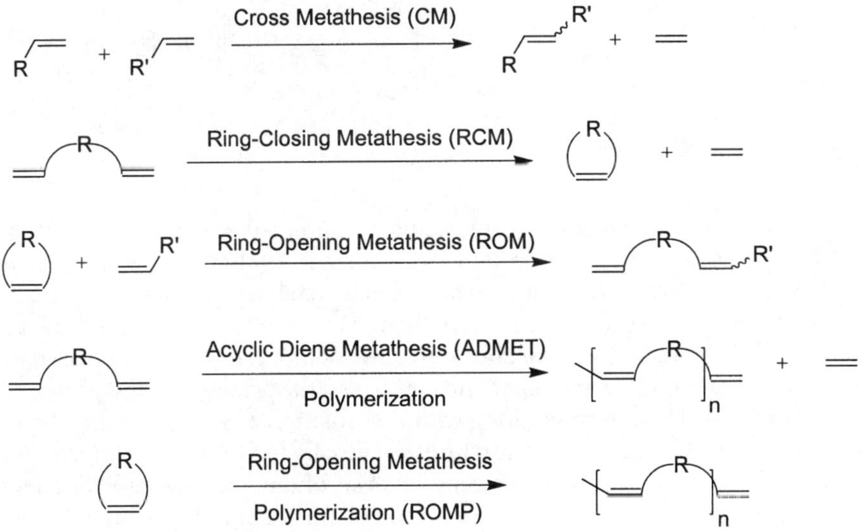

Scheme 1 Olefin metathesis

A metathesis reaction is defined as a chemical transformation in which atoms from different functional groups interchange with one another, resulting in the redistribution of functionality yielding similar bonding patterns for both molecules [6].

For olefin metathesis, two carbon-carbon double bonds are reacted to form two new olefins (Scheme 1). This transformation was initially reported in the 1950s, but it was not until 1967 that Calderon coined the term "olefin metathesis" [7–11]. Since then, olefin metathesis has been applied to polymer and small molecule synthesis. Pharmaceutical chemists rely on olefin metathesis to create complex cyclic systems, and previously difficult medium and large ring closures can now be achieved rather easily using ring-closing metathesis (RCM) [1, 2, 12, 13]. While RCM is performed at low concentrations to inhibit dimerization, reactions in the presence of high olefin concentrations yield polymers. Macromolecular chemists have embraced olefin metathesis, as it allows the preparation of functionalized hydrocarbon polymers through ring-opening metathesis polymerization (ROMP) [6, 14, 15] and acyclic diene metathesis (ADMET) [16–20].

1.1
History of Olefin Metathesis

Olefin metathesis began as an industrial process involving ill-defined heterogeneous catalysts comprising high oxidation state metal salts and various activating metal oxides [3]. Due to low concentrations of the active species, no spectroscopic evidence could be obtained and little mechanistic data was avail-

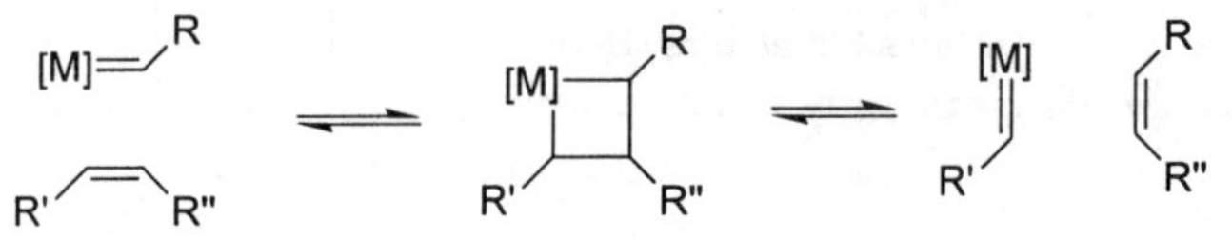

Fig. 1 Chauvin mechanism

able. Debates over the metathesis mechanism continued until the introduction of the now widely accepted Chauvin mechanism in 1971 (Fig. 1) [21, 22].

His proposal involved a metal carbene and a metallocyclobutane intermediate and was the first proposed mechanism consistent with all experimental observations to date. Later, Grubbs and coworkers performed spectroscopic studies on reaction intermediates and confirmed the presence of the proposed metal carbene. These results, along with the isolation of various metal alkylidene complexes from reaction mixtures eventually led to the development of well-defined metal carbene-containing catalysts of tungsten and molybdenum [23–25] (Fig. 2). After decades of research on olefin metathesis polymerization, polymer chemists started to use these well-defined catalysts to create novel polymer structures, while the application of metathesis in small molecule chemistry was just beginning. These advances in the understanding of metathesis continued, but low catalyst stability greatly hindered extensive use of the reaction.

In particular, Schrock-type catalysts suffered from extreme moisture and air sensitivity because of the high oxidation state of the metal center, molybdenum. Due to the oxophilicity of the central atom, polar or protic functional groups coordinate to the metal center, poisoning the catalyst and rendering it inactive for metathesis. Since late transition metal complexes are typically more stable in the presence of a wide range of functionalities, research was focused on the creation of late transition metal carbene complexes for use as metathesis catalysts.

Grubbs' first well-defined ruthenium carbene catalyst ([Ru]) was introduced in the early 1990s as the first air stable metathesis catalyst allowing for manip-

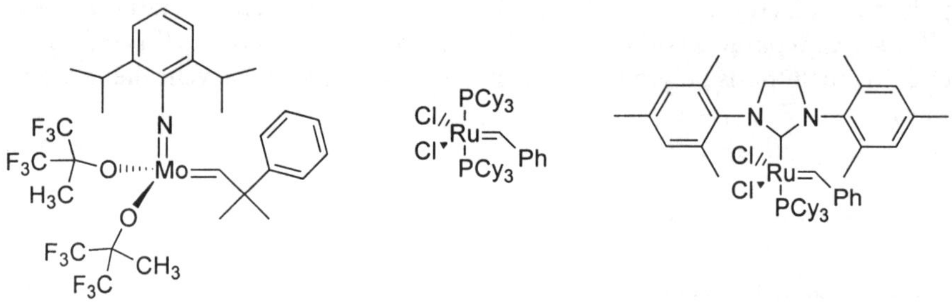

Fig. 2 Metathesis catalysts

ulation of this species outside of inert atmospheres (Fig. 3) [26]. Although catalyst stability was significantly improved, the metathesis activity was reduced substantially relative to Schrock-type catalysts, and the first-generation complexes exhibited slower, lower-yielding reactions. Metathesis activity and functional group tolerance were substantially increased for [Ru]* with the introduction of the N-heterocyclic carbene ligand as a replacement for one of the trialkyl phosphine ligands [27–36]. These catalysts exhibited activity close to that of [Mo], and this development brought olefin metathesis to the forefront of modern chemistry. However, one major disadvantage was later discovered; the second-generation complex simultaneously catalyzed metathesis and olefin isomerization. Since then, cross metathesis studies have revealed isomerization occurring at the same time as metathesis, leading to a myriad of olefin products [37, 38]. While this creates synthetic issues for the design of exact chemical structures through RCM or ADMET, polymer chemists who only desire the incorporation of functionality into polymer systems through ROMP or ADMET are still able to take advantage of the improved stability and reactivity of the [Ru]* complex.

1.2
History of ADMET

ADMET polymerization is performed on α,ω-dienes to produce strictly linear polymers with unsaturated polyethylene backbones, as shown in Scheme 2. This step-growth polymerization is a thermally neutral process driven by the release of a small molecule condensate, ethylene [16–20]. Ring-opening metathesis polymerization (ROMP) is widely used to polymerize cyclic olefins and is performed with the same catalysts as in ADMET polymerizations.

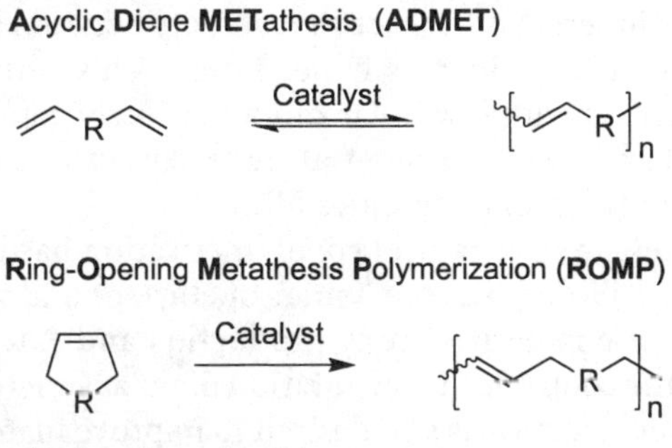

Scheme 2 Metathesis polymerization

The equilibrium of the ADMET polymerization is forced towards high polymer by running bulk polymerizations under vacuum to remove ethylene. Working under bulk conditions or in solution, ADMET polymer products have been isolated up to 80 kg mol^{-1} using [Mo] on hydrocarbon monomers and up to 70 kg mol^{-1} using [Ru]* on peptide functionalized monomers [39, 40].

Near-quantitative conversion of monomer to polymer is standard in these polymerizations, as few side reactions occur other than a small amount of cyclic formation common in all polycondensation chemistry [41]. ADMET depolymerization also occurs when unsaturated olefins are exposed to pressures of ethylene gas [42, 43]. In this case, the equilibrium nature of metathesis is shifted towards low molecular weight products under saturation with ethylene. Due to the high catalytic activity of [Ru]* and the ability of [Mo] and [Ru] to create exact structures, ADMET has proven a valuable tool for production of novel polymer structures for material applications as well as model copolymer systems to help elucidate fundamental structure property relationships [5].

2
Functionalized Polyethylene via ADMET: Model Copolymers of Ethylene and Vinyl Monomers

Polyethylene is the polymer produced in the greatest amounts (by weight) around the world, and is sought after for various applications due to its cost-effectiveness, ease of production and range of available polymer properties. For decades, researchers have been trying to produce functionalized polyethylenes in an attempt to enhance overall properties of the material through incorporation of polar groups along the polymer backbone. Addition of polar functional groups within the hydrophobic material has been shown to improve polymer adhesion, barrier properties, and chemical resistance [44].

Post-polymerization functionalization has been used to this end, but most research has been directed toward the copolymerization of ethylene with polar monomers. In this manner, inexpensive monomers can be used to create novel polymeric materials with a wide range of applications. The major drawback to this methodology is the inherent difference in reactivity between ethylene and other vinyl monomers during chain polymerization. This phenomenon is known to yield copolymers with low polar monomer incorporation and increased branch content arising from chain transfer events caused by side reactions with polar and/or protic functionalities [45].

Historically, high-pressure free radical copolymerization has been used to produce highly branched, ill-defined copolymers of ethylene and various polar monomers. Although these materials are in production and extensively used throughout the world, the controlled incorporation of polar functionality coupled with linear polymer structure is still desired to improve material properties. Recent focus in this area has led to the development of new transition metal catalysts for ethylene copolymerization; however, due to the electrophilicity of the metal centers in these catalysts, polar functional groups often coordinate with the metal center, effectively poisoning the catalyst. There has been some success, but comonomer incorporation is hard to control, leading to end-functionalized, branched polyethylenes [44, 46]. These results are undesirable due to low incorporation of polar monomer into the polymer as well as the inability to control branching, that leads to decreased material properties relative to the linear systems. Linear systems afford better materials due to their regular polymer structures, allowing for greater overall crystallinity. Since the goal of this research is to enhance polymer properties, incorporation of polar groups into a linear polyethylene backbone would be ideal, as material properties tend to increase with polymer crystallinity.

ADMET offers a synthetic route to strictly linear, functionalized polyethylenes through the polymerization of α,ω-dienes followed by exhaustive hydrogenation. Researchers have been able to use metathesis catalysts in conjunction with the functionalized monomers to produce statistical or sequenced copolymers of ethylene with various polar monomers. With the improved tolerance and reactivity of [Ru]*, the broadening of ADMET methodology will allow the syntheses of numerous functionalized systems [4]. However, due to the well known olefin isomerization that occurs during the metathesis polymerization with [Ru]*, monomer sequence control is lost and the methylene run length between functional groups varies widely.

While ROMP provides a facile synthetic route to functionalized polyethylene through polymerization and hydrogenation, the products lack monomer sequence control and usually fall within a short range of comonomer compositions relative to ethylene copolymers. This can be overcome by copolymerization, but as with any chain addition chemistry, the reactivity ratios of the two monomers – in this case olefins – must be essentially identical to obtain a truly random copolymer. Otherwise, a gradient polymer is obtained. ADMET does allow sequence control through the use of symmetrical α,ω-dienes affording vinyl monomer analog polymers, and due to the step-growth nature of the reaction, ADMET copolymerizations are truly random. These benefits of ADMET can be used to create ethylene copolymer analogs for materials testing and fundamental studies of polymer structure-property relationships. As illustrated in ethylene-propylene ADMET copolymer research, sequence control of comonomers in functionalized ethylenes results in higher degrees of crystallinity, leading to enhanced polymer properties [39, 47].

2.1
Ethylene-Propylene Copolymers

Controlling branching in polyethylene has been of significant synthetic interest for over 60 years, and numerous studies have been conducted on polyethylenes with different branch contents and structures [48–56]. Ethylene-propylene (EP) copolymers are methyl branched polyethylenes and can be used to study fundamental branching effects on polymer properties. Due to the large range of material properties available from branched polyethylenes, various architectures have been synthesized using free-radical, Ziegler-Natta, homogeneous metallocene, and, more recently, late transition metal catalysts [57–64]. However, side reactions such as chain transfer and chain walking persist in these polymerizations, causing unwanted branching and broad molecular weight distributions [65–67]. These structural defects, although exploited to create novel polyethylene materials, become unfavorable when trying to understand specific interactions and how they affect the microstructure in copolymers of ethylene and α-olefins. After decades of research focused on chain polymerization synthesis of strictly linear polyethylenes, Wagener et al. proposed a new synthetic approach to these polymers through step-growth metathesis polymerization, better known as ADMET.

Polyethylene modeling using ADMET step-polymerization began with the production of strictly linear polyethylene by polymerization of 1,9-decadiene followed by exhaustive hydrogenation (Scheme 3) [68, 69].

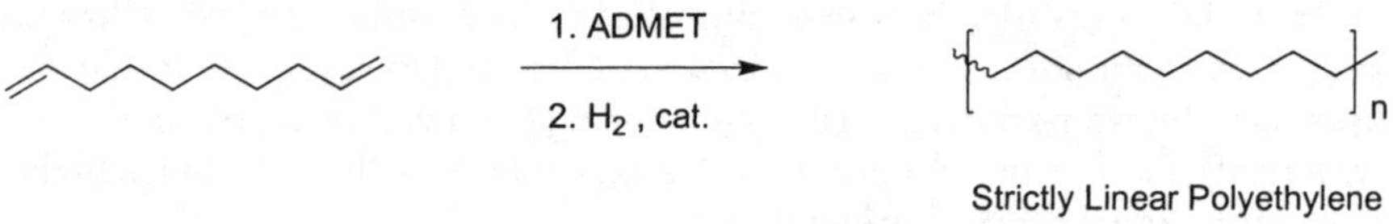

$$\text{1. ADMET} \\ \text{2. } H_2\text{, cat.}$$

Strictly Linear Polyethylene

Scheme 3 Linear polyethylene via ADMET polymerization

This metathesis polymerization successfully creates defect-free macromolecules due to the mild nature of metathesis and its lack of side reactions. Although the molecular weights of ADMET polymers are less than that of industrial polyethylene, the entanglement molecular weight for linear polyethylene is 1.0 kg mol^{-1}. Since the molecular weights of ADMET polymers are far above this value, polymers produced via ADMET make good model polymers. We now have the ability to fully control branching in linear polyethylene, allowing isolation of fundamental structure-property interactions.

Synthesis of precisely methyl-branched polyethylene began with the preparation of symmetrical methyl-branched α,ω-dienes to serve as ADMET monomers [71]. A family of monomers was designed not only to determine the effects of regular branching, but also to probe the effect of comonomer incorporation, in this case by adding propylene into an ethylene backbone. These molecules can be polymerized using [Ru] and later hydrogenated, affording the unsaturated, methyl-branched polyethylenes with number average molecular weights ranging from 8.5–17.5 kg mol^{-1} and PDIs approaching 2.0, typical of step-growth polymerization [73]. A family of polymers with multiple branch contents was obtained by varying the methylene run length between olefins in the monomer, leading to different spacings of the methyl branches along the polymer (Scheme 4).

For catalyst comparison, the same monomer was polymerized with [Mo] and [Ru], affording polymers of equal branch content and M_n values of 72.0 and 17.4 kg mol^{-1}, respectively. Upon thermal analysis of these two polymers, both produced sharp melt transitions at 57 °C, indicating no difference in polymer morphology across this range of molecular weights. In-depth discussions on structural and thermal characterization are included in this report [71]. These synthetic studies proved that ADMET step-growth chemistry was a viable

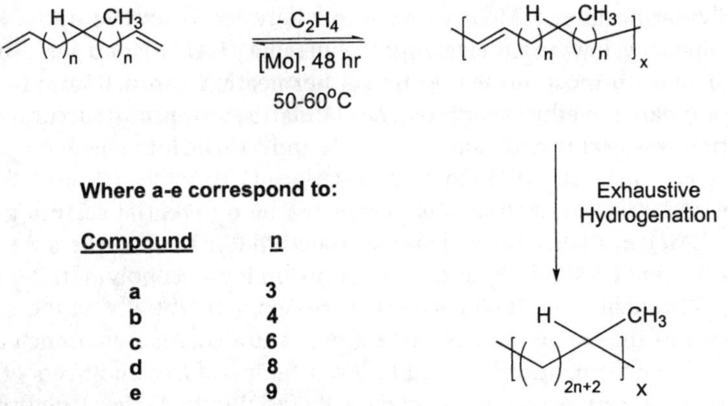

Scheme 4 Precise methyl branching in polyethylene (from [73])

alternative to chain addition polymerization for creating model polyethylenes. Control over molecular weight, and most importantly over branching and polydispersity, allow the kind of precise control over the polymer microstructure that has been sought for over 60 years.

Upon realizing that ADMET could be successfully used in a polymer modeling motif, expansion of ADMET methodology allowed for the modeling of random methyl branches by copolymerizing previously-used monomers with different weight percentages of 1,9-decadiene using [Mo] as the catalyst (Scheme 5). These experiments produced a family of randomly branched, linear polyethylenes containing 1.5–97.4 methyl branches per 1000 carbons in

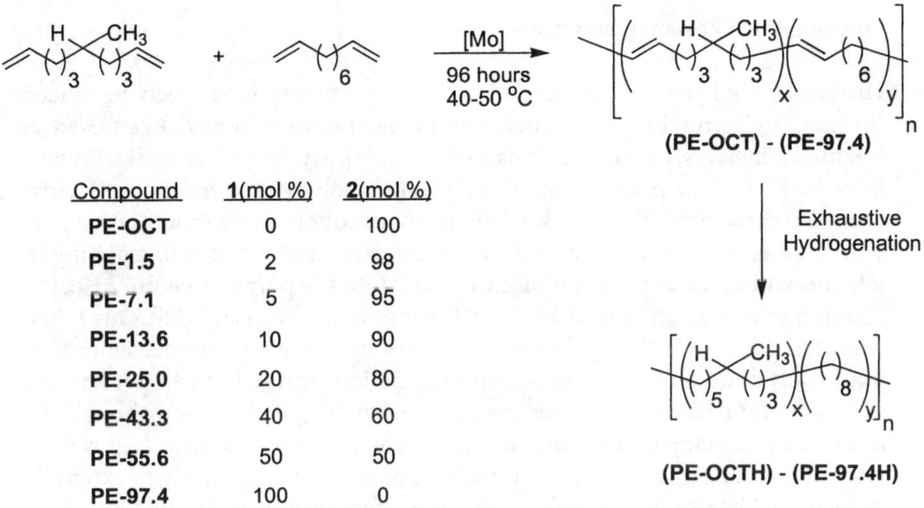

Compound	1(mol %)	2(mol %)
PE-OCT	0	100
PE-1.5	2	98
PE-7.1	5	95
PE-13.6	10	90
PE-25.0	20	80
PE-43.3	40	60
PE-55.6	50	50
PE-97.4	100	0

Scheme 5 Random methyl branching in polyethylene (from [72])

the polymer backbone. Molecular weights between 14 and 31 kg mol^{-1} were determined by low angle laser light scattering (LALLS) and were in close agreement with those measured by gel permeation chromatography (GPC) when compared to ethylene-propylene standards. Stringent structural characterization was performed using FT-IR, ^{1}H and ^{13}C NMR to check the random structures of these ADMET copolymerizations [72].

Thermal characterization was performed by differential scanning calorimetry (DSC) on all unsaturated and saturated ADMET copolymers. As expected, the percent crystallinity and heats of fusion for all copolymers containing up to 25 branches per 1000 carbons increased substantially as the polymer backbone of the copolymers became more saturated. Also, as branch content increased, melt transitions shifted to lower temperatures and heats of fusion decreased, indicating a reduction of overall crystallinity. As the structural order of the polymer chain is increased by hydrogenating the remaining olefins and reducing the branch content, the material is able to crystallize more easily, yielding higher melt transitions and improved material properties relative to industrial polymers. For highly branched systems with 43 branches per 1000 carbons or higher, broad endotherms were observed, similar to the broad melts seen with commercial polyethylenes. Even though ADMET copolymers represent a good model for ethylene-propylene copolymers over a wide range of branch contents, their lack of both extensive long chain branching and improved structural order is evident through characterization. Due to the growing interest in ADMET polymer modeling, current work includes the incorporation of longer branches into precisely and randomly branched systems to perfect the modeling and further understand industrial polyethylene materials.

2.2
Ethylene-Vinyl Acetate Copolymers

Historically, ethylene-vinyl acetate (EVA) copolymers have been produced through high-pressure free-radical copolymerization, and have been used in hot melt adhesives, packaging films, and toys [74]. Although free-radical chemistry has failed to produce many ethylene copolymers, ethylene and vinyl acetate represent an ideal, or Bernoulian chain copolymerization. In this particular case, the reactivity ratio product $r1.r2\cong1.0$, and $r1\cong r2\cong1.0$, resulting in a truly random incorporation of monomers into the polymer chain. Also, the copolymer composition is identical to the monomer feed ratio [75]. Due to the facile synthesis of EVA copolymers with varying degrees of vinyl acetate content, many studies on structure-property relationships have been performed, gathering fundamental data for polymer behavior [76–78]. Even though the availability of many EVA copolymers has created a large database, free-radical polymerization stills suffer from unwanted chain transfer, leading to extensive branching. Metathesis step polymerization offers a route to strictly linear EVA copolymers with a wide range of comonomer incorporation.

ADMET and ROMP followed by hydrogenation have been used to create novel, linear EVA copolymers (Scheme 6) [79, 80]. ROMP polymers were obtained by polymerizing a functionalized cyclooctene followed by hydrogenation using [Ru] residue under hydrogen pressure. Although these polymers were the first examples of linear EVA copolymers, ADMET methodology could expand our understanding of EVA copolymers, yielding sequenced copolymers that allow insight into fundamental polymer interactions.

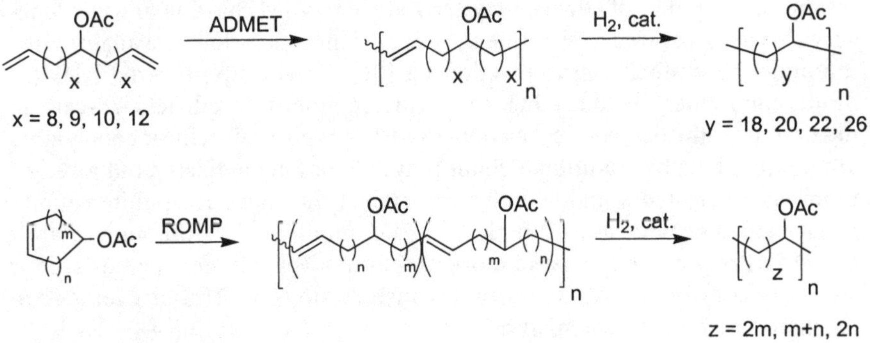

Scheme 6 EVA polymer modeling via metathesis polymerization

Starting with various acetate-functionalized dienes, ADMET polymerization was used to model sequenced EVA copolymers. These polymers were produced by bulk polymerization of the acetate-functionalized diene with [Ru] followed by hydrogenation (see Scheme 7). The hydrogenation procedure, reported as the saturation method for ROMP EVA copolymers, involves addition of silica to the reaction mixture after polymerization followed by addition of toluene. This heterogeneous catalyst mixture was then subjected to hydrogen pressures until the polymers were fully saturated. After filtration and solvent removal, colorless semicrystalline polymers of molecular weights ranging from 31–66 kg mol^{-1} were obtained. Due to the sequence control and lack of branching, narrow melting temperatures were found relative to the commercially-available Elvax series of EVA copolymers, usually exhibiting a broad melt transition typical of industrial copolymers. Experimental observations and characterization of ADMET model polymers illustrates the idea that synthesis of regular

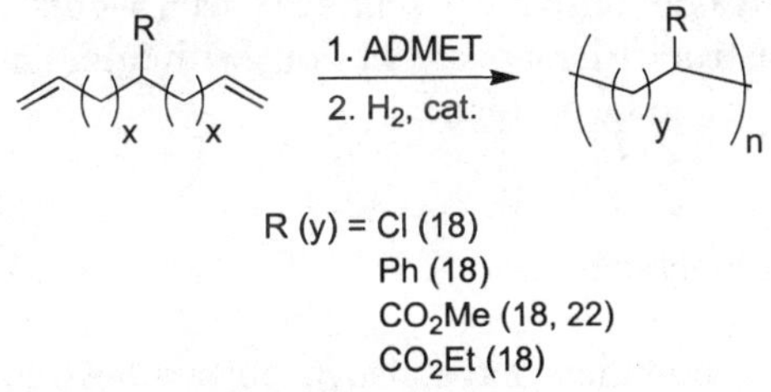

Scheme 7 Functionalized polyethylenes via ADMET

polymer structure allows access to novel polymer properties unattainable in current industrial processes.

2.3
Other Ethylene Copolymers

More industrial polyethylene copolymers were modeled using the same method of ADMET polymerization followed by hydrogenation using catalyst residue. Copolymers of ethylene-styrene, ethylene-vinyl chloride, and ethylene-acrylate were prepared to examine the effect of incorporation of available vinyl monomer feed stocks into polyethylene [81]. Previously prepared ADMET model copolymers include ethylene-*co*-carbon monoxide, ethylene-*co*-carbon dioxide, and ethylene-*co*-vinyl alcohol [82, 83]. In most cases, these copolymers are unattainable by traditional chain polymerization chemistry, but a recent report has revealed a highly active Ni catalyst that can successfully copolymerize ethylene with some functionalized monomers [84]. Although catalyst advances are proving more and more useful in novel polymer synthesis, poor structure control and reactivity ratio considerations are still problematic in chain polymerization chemistry.

2.3.1
Ethylene-Styrene Copolymers

Ethylene-styrene copolymers have been difficult to produce in the past due to reactivity differences between these monomers in chain polymerization chemistry. Although incorporation of styrene into a polyethylene backbone has been successfully achieved with Ziegler-Natta and metallocene catalysts, the styrene content is low and is usually included at the chain ends [85–88]. ADMET polymerization followed by hydrogenation has produced copolymers of this nature with exactly one phenyl substitution on every 19th carbon of polyethylene. Characterization by GPC yielded a molecular weight of 18 kg mol^{-1} when compared to polystyrene standards, while the presence of speculated low molecular weight oligomers and cyclics were detected at longer retention times. Thermal analysis of the styrene branched polyethylene using DSC showed two broad melt transitions at −22.5 °C and −1.5 °C. After annealing the sample between the two endotherms, a single melt transition at −6 °C is obtained with second order transitions detected as an unstable baseline between 5–20 °C. These interesting thermal characteristics are left as assumptions with future morphological and crystallographic work to be performed.

2.3.2
Ethylene-Acrylate Copolymers

Acrylate-ethylene copolymers have also been synthesized using the same tandem polymerization/hydrogenation methodology. Commercially, ethylene-methyl acrylate (EMA) and ethylene-ethyl acrylate (EEA) copolymers are available with about 20 wt% acrylate via high-pressure free radical polymerization. EMA and EEA copolymers suffer from extensive branching defects and large polydispersities; differences in the reactivity ratios of the two monomers make the synthesis of copolymers with varying comonomer contents difficult [89].

ADMET polymerization of methoxy- and ethoxycarbonyl-containing dienes followed by hydrogenation has been used to prepare suitable models of methyl acrylate and ethyl acrylate copolymers [83]. Both acrylate copolymers were synthesized with ester functionality on every 19th carbon. One methacrylate copolymer was synthesized with an ester on every 23rd carbon for direct comparison with similar EVA model copolymers already discussed in this review [79]. Molecular weights of 4.8 kg mol^{-1} and 6.4 kg mol^{-1} were obtained by GPC analysis for the methyl and ethyl acrylate polymers, respectively. Melting temperatures of 14.4 °C and 37 °C were found by DSC for the methyl acrylate polymers, again illustrating that an increase in functional group size on the polymer backbone leads to a greater decrease in polymer crystallinity. The ethyl acrylate polymer exhibited two melt transitions at 9.8 °C and 15.1 °C, similar to the phenyl-substituted polymer. Although the appearance of two endothermic events is not explained, the increased size of the pendant group is responsible for such a low melting point and possibly for second order transitions involving side group conformation similar to those seen in commercial EVAs and EMAs. The ethylene-methyl acrylate copolymer and the previously synthesized ethylene-vinyl acetate copolymer were compared in extensive discussions that also focused on NMR spectroscopy and thermal behavior [81]. This comparison is interesting, as EMA and EVA pendant groups exist as structural isomers.

2.3.3
Ethylene-Vinyl Chloride Copolymers

The final copolymer synthesized in this report on functionalized polyethylenes was an ethylene-vinyl chloride copolymer. Also referred to as chlorinated polyethylene (CPE), this copolymer is of significant interest to the polymer community for morphological and crystallinity studies [90–93]. Wegner et al. performed X-ray and DSC studies on CPEs developed from post-functionalization of polyethylene, which indicated that the relatively small chlorine atom can be incorporated into the polyethylene crystalline structure as a defect [92]. This means that CPEs could lead to materials with improved properties by adding polar functionality without sacrificing crystallinity. Wegner's studies proved that crystallinity control was dependent on functional group placement along the polymer chain. This is in agreement with all results obtained for the ADMET functionalized polyethylenes.

The synthesis of ADMET CPE started with the isolation of the symmetrical chlorine functionalized α,ω-diene. Upon polymerization with [Ru] followed

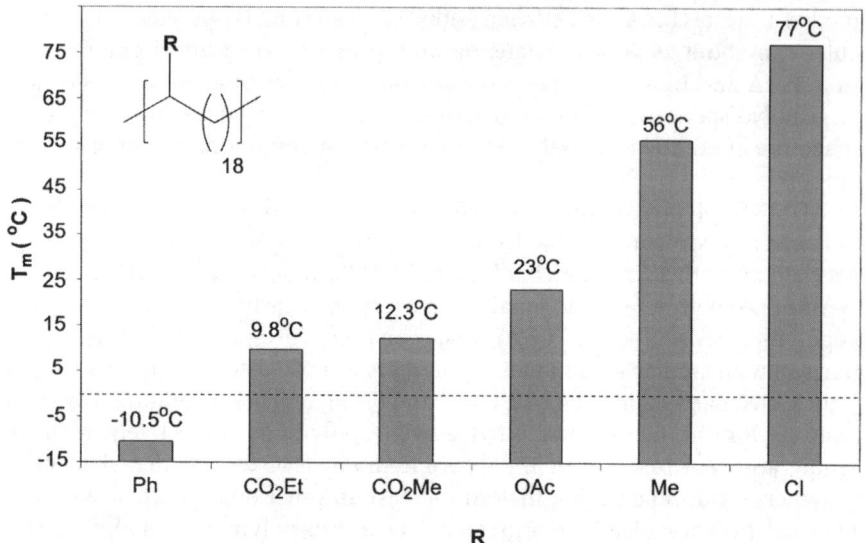

Fig. 3 Bar graph of melt transitions versus functional group on every 19th carbon (from [81])

by hydrogenation and precipitation in acidic methanol, a colorless crystalline polymer was obtained in good yield. Thermal analysis via DSC revealed an expectedly high melting point of 77 °C due to chlorine incorporation into the polyethylene crystal lattice. The melting point is sharp and well-defined, similar to melting points of precisely methyl-branched polyethylenes [71]. When comparing free radical ethylene-vinyl chloride and ethylene-propylene copolymers, a linear relationship is observed between melting points of the copolymers and the comonomer content [90]. For ADMET model copolymers, the situation is reversed, and the methyl and chloro polymers melt at 20 and 57 °C, respectively. In addition, the high degree of crystallinity and lack of branching accounts for the low solubilities of the model CPEs in various organic solvents.

A family of ADMET model copolymers were synthesized and used to study the effects of regular microstructure on polymer properties, in particular functionalized polyethylenes. The structure-property relationships of various ethylene copolymers can be clarified using these model systems. This is illustrated in Fig. 3 by the relationship of T_m to functional group size. Future studies on these and similar systems should lead to fundamental discoveries concerning the class of materials known as polyethylenes and their physical properties.

3
Block and Graft Copolymers via ADMET

3.1
Grafted Polyethylenes

ADMET has been used to create well-defined architectures of various functionalized polyethylenes that have been discussed throughout this review. Combinations of ADMET with atom transfer radical polymerization (ATRP) and ring-opening polymerization have both proven successful in synthesizing well-defined graft copolymers. Careful polymerization techniques are extremely important in grafted polymer synthesis as the graft density, graft length, and chemical nature of the graft must all be controlled to attain the desired polymer structure. Properties of grafted systems render them desirable for many purposes, including impact-resistant materials and polymeric emulsifiers, while poly(ethylene glycol) (PEG) grafted polymers have shown some success in biomedical applications due the biocompatibility of this polymer [94–99]. Applying macromonomer and macroinitiator polymerization techniques, ADMET polymerization in conjunction with other common polymerization techniques has afforded PEG and polystyrene (PS) grafted unsaturated polyethylenes [100, 101].

3.2
Polyethylene-g-Poly(Ethylene Glycol)

For this study, two polyether graft copolymers were synthesized, one containing a short polyether graft and another with an extended graft architecture (Scheme 8) [102]. The short graft was prepared by reacting an alcohol functionalized diene with methoxyethoxymethyl chloride (MEMCl) to afford the MEM functionalized alcohol. This monomer was polymerized with [Ru] affording the precisely spaced short graft copolymer. Although molecular weight determination by NMR endgroup analysis was impossible, GPC analysis by comparison to PS standards determined a molecular weight of 15 kg mol^{-1} for the MEM grafted copolymer.

For the longer polyether graft, the same alcohol functionalized diene was used to initiate a ring-opening polymerization of ethylene oxide. This reaction afforded a graft length of twelve repeat units of PEG on the diene macromonomer, which was verified via MALDI-TOF analysis, confirming the monodispersity of the graft length prior to polymerization. Upon isolation of the PEG grafted diene, polymerizations were conducted using [Ru], and a polymer of 5.5 kg mol^{-1} was obtained. Catalyst complexation with the PEG graft may be responsible for low conversions during the ADMET reaction. Macromonomer polymerization was also performed using [Ru]*, affording a 12.0 kg mol^{-1} polymer as analyzed by GPC. Although this report refers to these structures as "perfect comb" graft copolymers, it is now well known that [Ru]* isomerizes

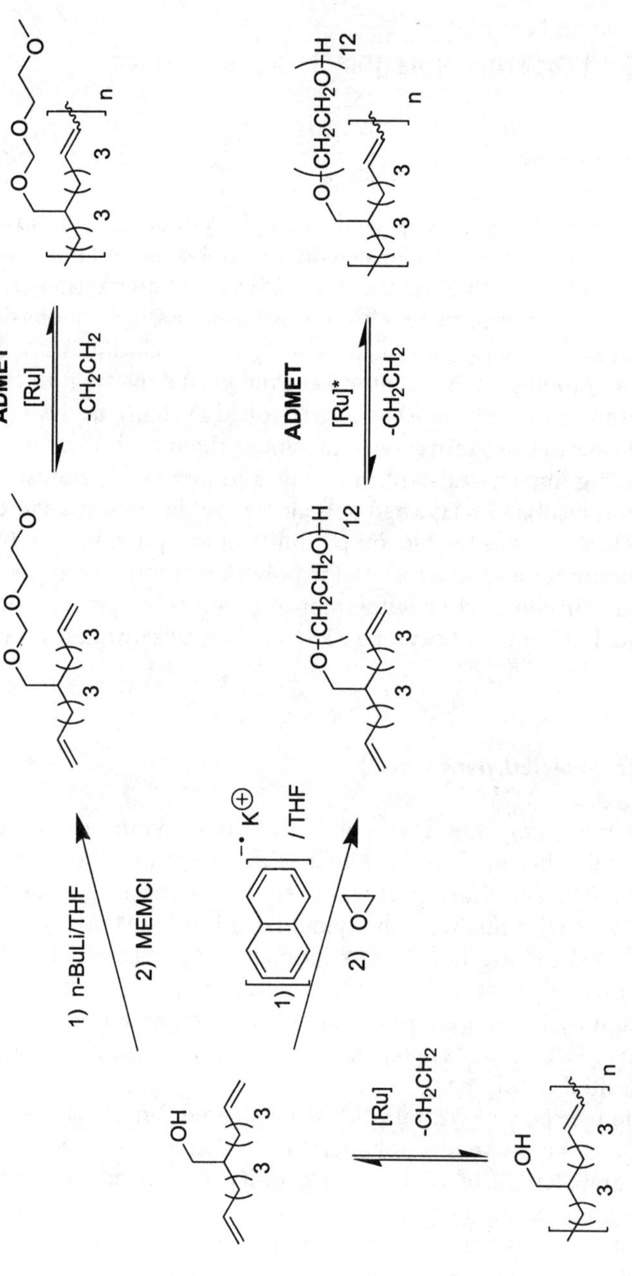

Scheme 8 Polyether grafted ADMET polymers (from [102])

olefins prior to and during polymerization, yielding polymers with irregular methylene run lengths between grafts. Despite imperfect polymer structure, a novel PEG grafted unsaturated polyethylene was successfully prepared. This copolymer demonstrated interesting thermal behavior when analyzed by DSC, and extensive discussion concerning thermal and spectroscopic characterization is included within the article [102].

3.3
Polyethylene-*g*-Polystyrene

Polyethylene-*g*-polystyrene has also been prepared using similar macromonomer and macroinitiator techniques [103]. In these experiments, ATRP was performed with an α,ω-diene macromonomer or a previously synthesized polymer was used as a multiple point macroinitiator (Scheme 9). For the ADMET macromonomer synthesis, PS grafts were prepared by exposing the diene initiator to ATRP conditions in the presence of styrene. Following removal of the initiator by passing the crude polymerization mixture through silica, two macromonomers with graft lengths of 15 and 30 repeat units were isolated by precipitation in cold methanol.

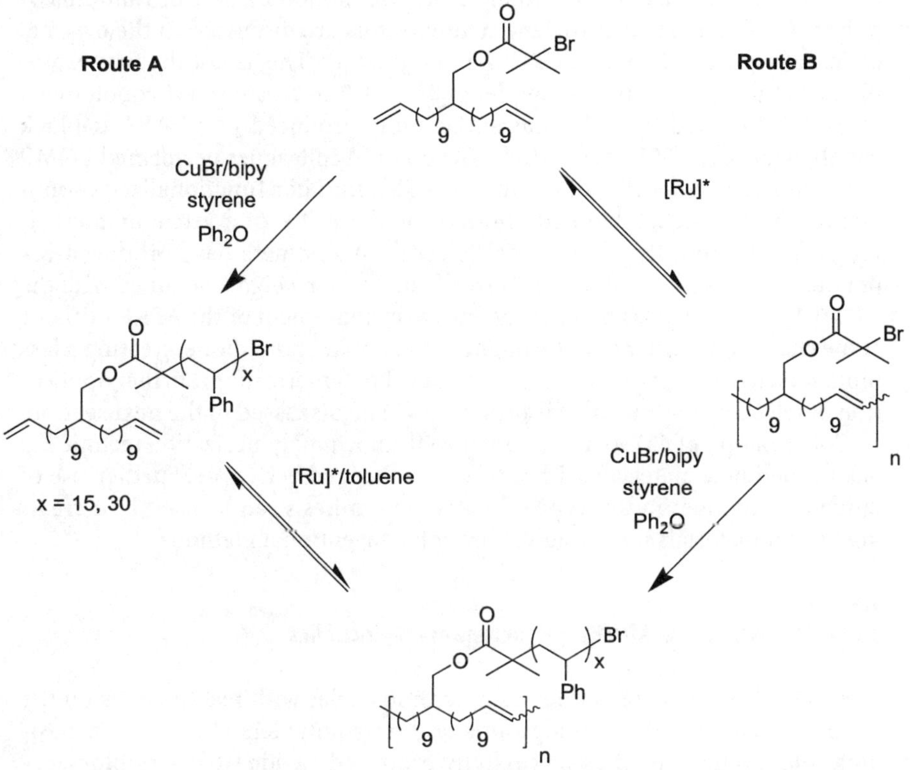

Scheme 9 Polystyrene grafted ADMET polymers (from [103])

All graft lengths were verified by MALDI-TOF analysis, and the molecular weights of two macromonomers were determined at 2–3 kg mol^{-1} by GPC. Since these macromonomers existed as solids, ADMET polymerization could not be performed in the bulk, so polymerizations were performed in solution. Dissolution of the macromonomer in dried, degassed toluene followed by addition of [Ru]* afforded only dimerization for the large PS-grafted macromonomer and a degree of polymerization of 5 for the shorter PS-grafted macromonomer. Steric hinderance was speculated as the cause for these low conversions as the large PS grafts reside close to the metathesis reactive site on the macromonomer. The PS-grafted diene was copolymerized with 1,9-decadiene, but similar conversions were obtained and the polymers exhibited number average molecular weights between 7.4 and 13.8 kg mol^{-1}. Again, macromonomer sterics were to blame for the low olefin conversion. Once macromonomer techniques were exhausted, the ATRP initiator functionalized diene was polymerized yielding the highest molecular weight polymer in this study at 17 kg mol^{-1}. Application of this polymer as a macroinitiator was attempted under ATRP conditions in the presence of styrene; however, only short graft lengths were obtained.

The thermal characterization of these novel grafted systems revealed unique behavior not associated with either of the two homopolymers or random copolymers of styrene and ethylene. These results are discussed in the original article along with NMR and GPC characterization. This is not the first example of the use of metathesis coupled with ATRP to create novel copolymers, as ROMP followed by ATRP has successfully produced novel ABA triblock copolymers [104, 105]. In this study, Grubbs and colleagues performed ROMP on cyclooctene with a small amount of ATRP initiator functionalized monoolefin. The telechelic obtained allowed for the ATRP of styrene or methylmethacrylate from the chain ends. Telechelic monomers have produced copolymers up to 40 kg mol^{-1}, well above the molecular weights obtained with the ADMET studies. This is most likely due to the placement of the ATRP initiator at the chain ends rather than throughout the polymer backbone, creating a less hindered environment for subsequent radical polymerization. Further application of telechelics as metathesis products will be discussed in the next section.

Combination of ADMET with other well-known polymerization techniques has yielded new materials with interesting thermal behaviors. Further use of tandem techniques in this type of materials synthesis can be useful for creating hybrid materials unattainable through conventional methods.

3.4
Block Copolymers via ADMET Polyoctenamer Telechelics

Telechelic polymers are defined as macromolecules with reactive sites on the polymer chain, usually as endgroups on linear polymers [106]. This macromolecular architecture has successfully produced a wide variety of block copolymers using macroinititated polymerizations. Living anionic polymerizations have been used to create well-defined block copolymers for various applications. In this case, the anionic chain ends act as initiators in the sequential polymerization of vinyl monomers. This method allows for the preparation of highly monodisperse block copolymers that exhibit a range of widely varying physical properties. Upon termination of the anionic chain ends with elec trophilic reagents such as anhydrides, carbon dioxide, or ethylene oxide, selective placement of the desired end groups can be achieved for future functionalization or polymerization. Initiators containing protected functionality have also been used to this end when, upon deprotection, reactive groups reside on chain ends. Diblock copolymers are synthesized through this methodology using the macroinitiator approach to polymerize a different monomer from the reactive chain end. For metathesis polymerization, monofunctional olefins are added to the polymerization mixture not only to control overall molecular weight, but also to attach desired functionality to the chain ends.

A variety of telechelic polymers have been synthesized through tandem ADMET and cross metathesis experiments [107–109]. Although the vinyl end groups created from ADMET polymerization are reactive and able to be used in subsequent polymerization, chemical differentiation of these olefins from the internal olefins in the polymer chain is difficult in most reactions and impossible in metathesis chemistry. Copolymerization of an α,ω-diene monomer with a previously synthesized ADMET polymer yielded a copolymer randomized through *trans*-metathesis along the polymer backbone [110, 111]. The inability to differentiate olefins leads to a randomization of olefins within the metathesis regime, and end groups can be incorporated into the polymer using a functionalized monoolefin that can undergo cross metathesis with any double bonds in the polymer creating chain ends in situ. As long as the functionality within the endcapping unit does not affect metathesis, telechelic polymers of desired mass and low polydispersities can be prepared. Recently, triblock copolymers and segmented copolymers have been synthesized using this methodology. Subsequent polymerization from the chain ends and use of the telechelic as a reactive monomer in a second polymerization has been performed, yielding novel block copolymer architectures.

Brzeznska and Deming have synthesized novel poly(γ-benzyl-L-glutamate)-*b*-polyoctenamer-*b*-(γ-benzyl-L-glutamate) triblock copolymers (Scheme 10) [112]. These copolymers are of interest as the incorporation of biopolymers into a block system may offer morphological control through tunable conformations of biological materials. Specifically, the production of synthetic protein-lipid bilayers for membrane development has been targeted [113]. Previous work has shown that influences of alpha and beta conformations of biological systems can alter the morphologies and thermal properties of common homopolymers [109, 114]. The recent development of low glass transition telechelic polymers has become of interest in regard to the development of novel polymeric networks. As the methodology of incorporating chemically different polymers into covalent networks becomes available, access to previously unattainable morphologies and polymer properties will also become available.

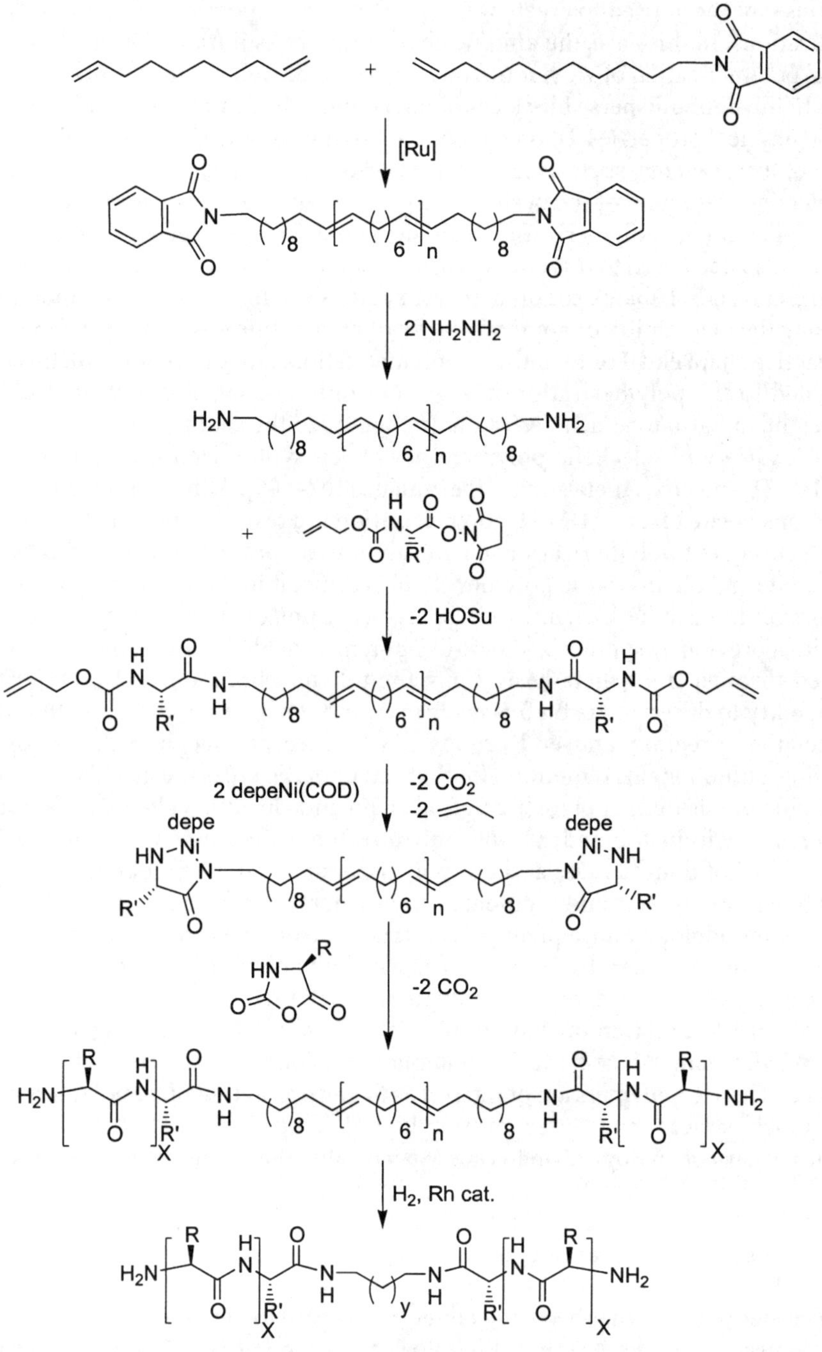

Scheme 10 Block copolymers via ADMET telechelics (adapted from [112])

Synthesis of poly(γ-benzyl-L-glutamate)-*b*-polyoctenamer-*b*-(γ-benzyl-L-glutamate) triblock copolymers began with the preparation of a telechelic polyoctenamer via ADMET polymerization [112]. 1,9-decadine was polymerized in the presence of 11-phthalimido-1-undecene with [Ru], affording a telechelic polymer endcapped with phthalimido groups. Incorporation of the phthalimido groups undoubtedly begins with the polymerization, but trans metathesis incorporates units of 1,9-decadiene while retaining chain end functionality. In this case, telechelics were synthesized at desired molecular weights of 2.1–12.5 kg mol^{-1} by careful control of catalyst to monomer ratio. Molecular weight data were obtained by endgroup analysis in ^{1}H NMR and by GPC. Upon full characterization of the telechelic, the phthalimido endgroups were converted to amino end groups by reaction with hydrazine. This deprotected telechelic was then utilized as a macroinitiator in the living polymerization of glutamatic acid-*N*-carboxyanhydride [115, 116, 117]. Thermal characterization by DSC isolated melt transitions arising from the polyoctenamer block at 48.6 °C. Selective polyoctenamer degradation studies using osmium tetroxide allowed isolation of homo poly(glutamic acid) and confirmed the identity and structure of different blocks within this copolymer [119]. Upon isolation of the unsaturated triblock copolymer, hydrogenation was performed with Wilkinson's Rh catalyst to attain a linear polyethylene middle block, in the hope of raising the melt transition of this block, affording desirable material properties. Creation of homopolypeptide contaminants was avoided through the use of macroinitiator methodology that only allowed initiation from the polyoctenamer chain ends. This work has proven beneficial, illustrating the flexibility of ADMET chemistry in syntheses of novel materials with possible biological activity.

Tamura and colleagues have synthesized novel triblock copolymers through a starting middle block of epoxy functionalized polyoctenamer (Scheme 11) [119]. A telechelic polyoctenamer was synthesized via metathesis condensation of 1,9-decadiene with an epoxy functionalized monoolefin at 90 °C for 96 hours. A relationship between molecular weight and temperature dependence was determined and allowed synthesis of the ideal telechelics for these experiments. Ethylene removal was accomplished through intermittent vacuum cycles every three hours. Following precipitation of the reaction mixture from chloroform into acetone to remove catalyst residue, NMR and GPC analysis confirmed the synthesis of the epoxy telechelic of M_n 4700. Yields of the telechelic synthesis were around 25%, but high conversion of the olefin was noted in crude NMR experiments. The authors attributed these results to the loss of low molecular weight oligomers during post polymerization precipitation.

Copolymerization of the telechelic polyoctenamer diepoxide was performed by exposure to toluene diisocyannate, tributyl phosphate, and lithium bromide in toluene. Upon reaction of the isocyannate and epoxide, an oxazolidone moiety is placed in-between each segment of polyoctenamer. IR, NMR, and GPC analysis support these claims, but molecular weight data could not be obtained. GPC data alluded to the presence of two polymeric species in the

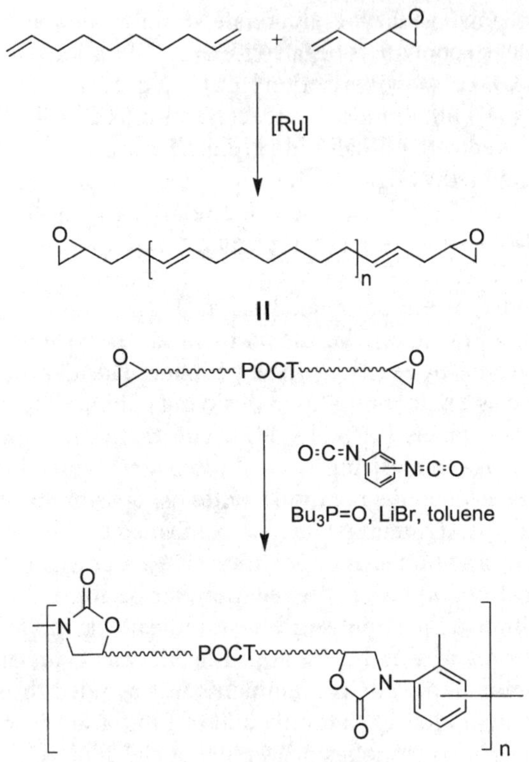

Scheme 11 Segemented copolymer via ADMET telechelic

product and current work is underway to separate and analyze the reaction products. Evaluation of small molecule test reactions of epoxy olefin dimerization and small molecule oxazolidone synthesis was performed and is also discussed within the article.

Numerous block copolymer architectures can be devised using functionalized polyoctenamer telechelics via ADMET polycondensation. Using protecting group chemistry or metathesis friendly functionalized monoolefins, almost any reactive group can be placed on the terminus of polyoctenamer through the previously reported procedures. One other example of silane-terminated telechelics was recently published, and discussion of this is included in the later section concerning silicon-containing polymers.

3.5
Alternating Copolymers

Alternating copolymers have been previously synthesized via metathesis polymerization by Grubbs et al. using ring-opening insertion metathesis polymerization (ROIMP) [120, 121]. Here, a fast ROMP polymerization of a cyclic olefin is followed by incorporation of a linear electron deficient diolefin into the polymer chain via in situ CM. While this method offers a synthetic route to AB copolymers, monomer selection is limited by ROIMP methodology. Recently, ADMET has been applied to this synthetic problem of affording novel AB alternating copolymers of widely varying structures.

Insertion metathesis polycondensation based on ADMET has been developed as a method of alternating copolymer production. Alternating diene metathesis polycondensation (ALTMET) was performed by reacting electron-poor dienes with terminal alkyl diolefins [122]. This polymerization successfully creates perfectly AB alternating structures due to the use of olefins with widely varying reactivity. Production of the electron-poor diene cross-product is suppressed due to the lower activity of the acrylate relative to the alkyl-substituted olefin and the inherent inactivity of the acrylate-functionalized catalyst intermediate towards other electron poor olefins. This selectivity allows for the polymerization of the alkyl diolefin followed by CM of the electron-poor olefin into the polymer backbone. A library of alternating copolymers was produced by copolymerization of monomers from families of alkyl diolefins and electron-poor diacrylates [122]. Since no electron-poor diolefin can be placed next to a similar monomer in the polymer, acrylates are incorporated into the polymer between two dissimilar monomers, creating the alternating AB copolymer. A wide range of unsaturated polyesters were produced, illustrating the flexibility of ALTMET relative to ROIMP for the synthesis of alternating copolymers.

4
Polymeric Materials via ADMET

4.1
Phosphazene Polymers

Phosphazenes are an interesting class of hybrid materials that have been used to create polymers with widely varying mechanical properties, and they have been recently incorporated into polyimides, polyamides, polyesters, polyurethanes, and polyketones [123–131]. Offering synthetic flexibility due to the labile P–Cl bond, parent cyclic phophazene trimers are easily accessible starting materials that allow substitution of almost any nucleophile to the phosphazene prior to polymerization. Linear phosphazene polymers have been produced in this manner with a wide range of substituents, allowing tunability of polymer properties [132]. Many of these polymers are suitable for use in fuel cell membranes, ion transport, elastomers, and biodegradable polymers. To date, literally hundreds of phosphazene polymers have been produced, ranging from water soluble, biodegradable polymers to hydrophobic elastomers [123, 124, 132]. Using metathesis to incorporate cyclic phosphazene units into commodity polymers like polyethylene has been of interest for morphology control as well as preparation of combustion-resistant materials.

Allcock and colleagues have used ROMP and ADMET to create novel polymeric systems containing cyclic phosphazene trimer unties in the polymer. Using ROMP, phosphazene units are incorporated as pendant groups throughout the main chain of an unsaturated polymer by polymerization of a functionalized norbornene [133]. For ADMET polymerization, a cyclic phosphazene was disubstituted with a terminal diene yielding the target monomer. Upon bulk polymerization, the cyclic phosphazene was included in the main chain of the polymer, creating a cyclolinear macromolecule [134–136]. As seen in Scheme 12, many cyclolinear polymers were prepared, containing various substituents.

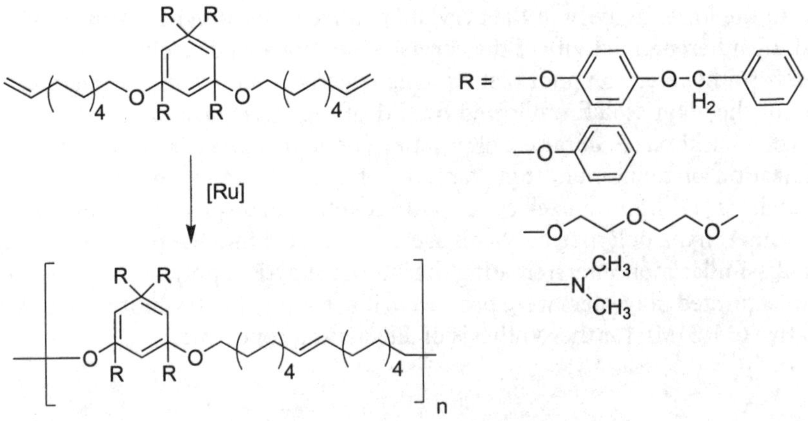

Scheme 12 Phosphazene-containing ADMET polymers

Allcock and colleagues isolated the off-white polymers by reprecipitation from THF into hexanes. Subsequent hydrogenation was performed, and as expected, the hydrogenated products exhibited higher melt transitions relative to the unsaturated polymers. An overall decrease in polymer melting point relative to polyethylene was observed due to the disruption of polymer crystallinity by incorporation of the cyclic moieties into the system. While the crystallinity decrease was expected, more robust materials were desired. To achieve this end, polymers containing more sterically encumbered groups and lower incorporations of the cyclic phosphazene were prepared and exhibited higher glass transition temperatures. Although material properties were significantly improved, preparation of polymers with even lower phosphazene contents were desired. Copolymerization of the phosphazene monomer with 1,9-decadiene afforded a copolymer of higher crystallinity and low phosphazene incorporation. Combustion resistance experiments are ongoing.

4.2
Poly(p-phenylene vinylene) Oligomers

Poly(p-phenylene vinylene) and various substituted derivatives belong to the family of conducting polymers [137]. This family of polymers can conduct electrons along the polymer chain through an uninterrupted arrangement of π orbitals. These polymers exhibit many unique properties due to this conductivity, and much research is underway to incorporate these flexible conductors into numerous applications. Due to low densities relative to metallic conductors and the chemical resistance of these systems, lightweight electronic devices can be fabricated and put into use where old technology fails. Electroluminescence, nonlinear optical response, photoconductivity, and photoluminescence are just a few properties of these polymers that make them attractive for use in many novel applications or devices [138–143]. Production of viable organic light emitting diodes is of interest, as applications of these polymers in flat panel displays become a reality.

Although much is known about these polymers, they continue to be studied throughout the world to delineate structure-property relationships, especially relating to conjugation length and electronic properties. Polymer modification via substitution has proven useful in determining systemic effects on bulk polymer systems, but additional research is directed toward conjugated oligomers to determine precisely what happens as you add repeat units into the system one at a time. These studies should yield a fundamental understanding of how conjugation length affects material properties. Once these relationships are well understood, novel conjugated polymers of desired conjugation lengths may be constructed, affording enhanced electronic properties for use in various devices.

Recently, a family of diheptyloxy-substituted PPV oligomers has been synthesized using ADMET methodology [144]. Upon polymerization of 2,5-deheptyloxy-1,4-divinylbenzene with [Mo], vinyl-endcapped PPV oligomers were obtained (Scheme 13). Although the vinyl-endcapped PPVs lack thermal stability and cannot be used directly for any applications, the vinyl groups allow unique possibilities for further functionalization through cross metathesis of the oligomers. In this case, *trans*-3-hexene was added to the polymerization mixture to endcap the oligomers with 1-butenyl groups rather than the vinyl substituents. Column chromatography with basic alumina was used to separate the vinyl and butenyl oligomers into crude mixtures, followed by subsequent rounds of chromatography on the crude mixtures to isolate the single oligomers of each type.

Characterization of the oligomers was performed using DSC, MALDI-TOF, FT-IR and NMR. DSC analysis on the monomer and all endcapped oligomers revealed an expected trend of increasing melting point with increasing oligomer molecular weight. MALDI-TOF was used to verify the monodispersity of each oligomeric sample, ensuring that chemical separation was suitable for isolating single oligomers. Extensive characterization by FT-IR and NMR

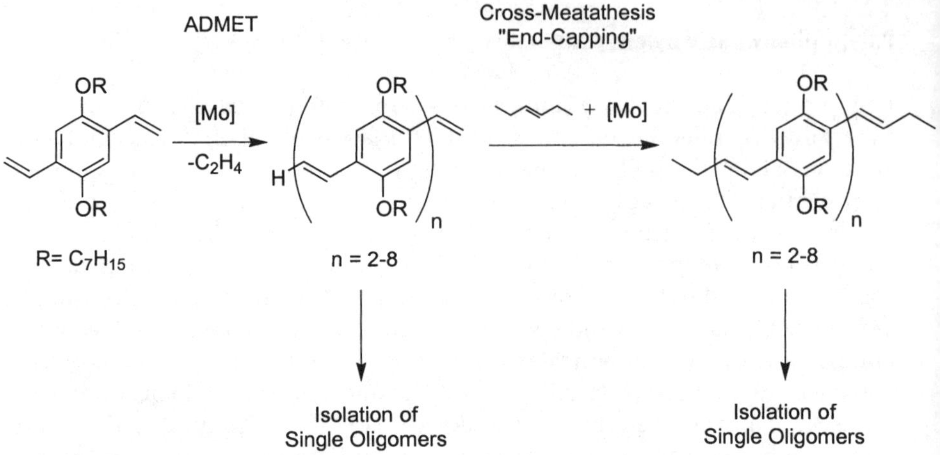

Scheme 13 PPV oligomers (adapted from [144])

was performed and is included within the text, focusing on the relation of oligomer size to spectral properties [144].

Upon isolation and characterization of single oligomers containing 2–8 PPV repeat units, research was focused on preparation of easily separable, higher order oligomers through metathesis telomerization; the polymerization of low oligomers [145]. Using [Mo], selective telomerization was performed on monodisperse low oligomers to obtain only CM products with terminal double bonds and no reaction with internal vinylene units. Since only the terminal double bonds of monodisperse oligomers react, product mixtures will contain oligomers of integer multiplicity relative to the starting material. This creates oligomeric mixtures differing by the number of repeat units in the original low oligomer material. Easily separable reaction mixtures of all *trans* configured PPV oligomers can be isolated and purified on a large scale using standard chromatographic techniques. Application of previously described methods fails in this motif if higher degrees of polymerization or larger quantities of monodisperse oligomers are needed. NMR analysis of products and time-dependent MALDI-TOF characterization of telomerization reactions are discussed in detail. While [Ru]* seems the catalyst of choice for dimerization of low oligomers, [Mo] is the catalyst of choice for telomerization due to higher conversions, no "side reactions" with internal vinylene units, and the kinetic control displayed during telomerization. Selective dimerization of lower oligomers has been shown to be a valuable method of monodisperse *trans*-PPV oligomer synthesis.

The PPV oligomers synthesized here represent a novel class of oligomers that prove valuable as model compounds for furthering the understanding of conjugated systems. Optical characterization along with detailed microstructure analysis of the oligomers correlated to the chain length should yield valuable data concerning conjugation length and optical properties of conjugated polymers.

5
Chiral Polymers via ADMET

Chiral polymers have been applied in many areas of research, including chiral separation of organic molecules, asymmetric induction in organic synthesis, and wave guiding in non-linear optics [146, 147]. Two distinct classes of polymers represent these optically active materials: those with induced chirality based on the catalyst and polymerization mechanism and those produced from chiral monomers. Achiral monomers like propylene have been polymerized stereoselectively using chiral initiators or catalysts yielding isotactic, helical polymers [148–150]. On the other hand, polymerization of chiral monomers such as diepoxides, dimethacrylates, diisocyanides, and vinyl ethers yields chiral polymers by incorporation of chirality into the main chain of the polymer or as a pedant side group [151–155]. A number of chiral metathesis catalysts have been made, and they have proven useful in asymmetric RCM as well as in stereospecific polymerization of norbornene and norbornadiene [156–159]. This section of the review will focus on the ADMET polymerization of chiral monomers as a method of chiral polymer synthesis.

5.1
Amino Acid-Containing Polymers

Amino acid-containing polymers are target molecules of great interest due to their possible application as biocompatible materials. Incorporation of glutamic acid functionality into a triblock copolymer has already been discussed in a previous section of this review [116]. Incorporation of these moieties into commodity polymers resulting in chiral materials is now feasible due to recent advances in amino acid isolation and purification [160]. Polymer chemists are interested in amino acids due to the hydrogen bonding nature of these units and their ability to take on a higher degree of order through α-helix or β-sheet formation. These ordered structures may lead to enhanced polymer behavior or advanced applications, as found in drug-delivery agents, chiral recognition stationary phases, or metal-ion absorbents [160].

Synthetic methods targeting amino acid incorporation into functional materials vary widely. Free-radical polymerization of various amino acid substituted acrylates has produced many hydrocarbon-amino acid materials [161, 162]. In separate efforts, Morcellet and Endo have synthesized and meticulously characterized a library of polymers using this chain addition chemistry [163–166]. Grubbs has shown ROMP to be successful in this motif, polymerizing amino acid substituted norbornenes [167–168]. To remain within the scope of this review, the next section will focus only on ADMET polymerization as a method of amino acid and peptide incorporation into polyethylene-based polymers.

Wagener and colleagues first reported the use of ADMET as a method of amino acid incorporation into polymers in 2001, and have since expanded their research to focus on the production of novel polymeric materials targeted

Table 1 Properties of amino acid-containing ADMET polymers I (from [169])

Monomer	Polymer	$[\alpha]_m$ (deg)	$[\alpha]_p$ (deg)	M_n (g/mol)[a]	PDI[b]	T_m (°C)[c]	T_m (°C)[d]
4	12	−32	−32	4700	1.73	e	e
5	13	−34	−20	27000	1.77	29	38
6	14			33000	1.64	e	39
7	15	−13	−64	31000	2.02	114	114
8	16	−13	−7	26000	2.10	135	135
9	17			21000	1.70	e	46

[a] M_n values were calculated by GPC vs polystyrene standards. Specific rotations were measured in CH_2Cl_2 at 25 °C, where $[\alpha]_m$ is the specific rotation of the monomer and $[\alpha]_p$ is the specific rotation of the polymer; [b] the polydispersity was determined by GPC analysis; [c] determined by DSC. Reported T_m is due to melt crystallization; [d] Determined by DSC. Reported T_m is due to solution crystallization; [e] No T_m was detected.

at biological applications [169, 170]. The first report revealed a novel approach for synthesis of amino acid moieties within the main chain and as pendant groups in an unsaturated polyethylene backbone (Table 1, Schemes 14 and 15). Due to reaction mixture solidification after only a few couplings, standard bulk polymerization procedures were inappropriate, and solution polymerization procedures were developed to counteract this phenomenon. Polymers synthesized with amino acids within the main chain were hydrolytically unstable, as degradation of the amino acid lead to polymer degradation. These

Table 2 Properties of amino acid and peptide ADMET polymers (from [170])

Polymer	M_w (g/mol)[a]	PDI[a]	T_m (°C)[b]	T_g (°C)[b]
19	26,000	1.54	e	28
20	36,000	1.45	e	18
21	73,000	1.55	132	d
22	63,000	1.67	e	7
23	21,000	1.62	38[c]	−21
24	25,000	1.91	e	5
25	42,000	1.85	e	−10
26	29,000	1.59	e	3
27	44,000	1.80	79	d
28	26,000	1.40	e	69
29	73,000	1.50	106	5
30	21,000	1.40	71[c]	d
31	38,000	1.64	74	d

[a] M_w and PDI values were calculated by GPC using LALLS; [b] data obtained using a Perkin-Elmer DSC 7 at 10 °C/min; [c] the T_m reported is that of the solvent crystallized sample and no T_m was observed from the melt crystallized sample; [d] no T_g was observed over the scanned range of −80 to +180 °C; [e] No T_m was observed over the scanned range of −80 to +180 °C.

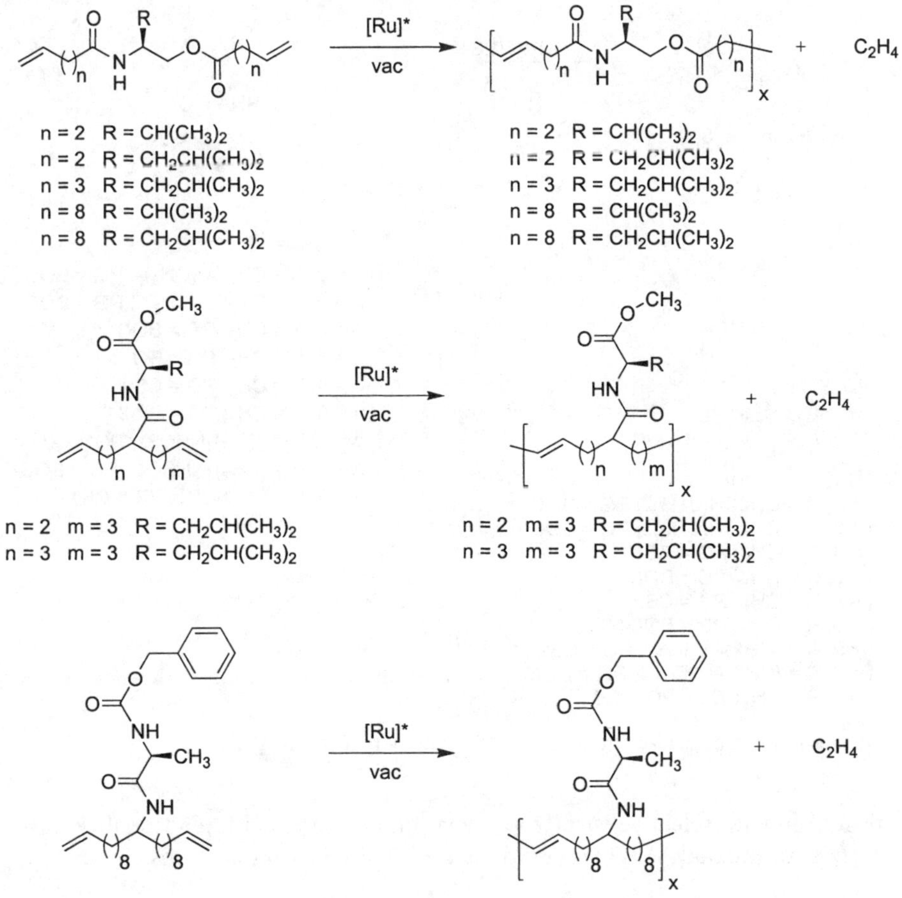

Scheme 14 Amino acid-containing ADMET polymers I (from [169])

polymers were synthesized in the hope that they could be used as biodegradable materials.

Pendant amino acid and peptide polymers synthesized via ADMET have been studied in greater detail than their linear analogs; to date more than twenty pendant functionalized unsaturated polyethylenes have been prepared and characterized (Table 2, Scheme 16) [170]. Synthesis of these polymers began with the synthesis of an amino acid functionalized diene for polymerization with [Ru]* in THF at 50 °C. Extra catalyst and solvent was added when necessary. Initial thermal characterization of unsaturated polymers revealed significant differences in morphologies, with some polymers being fully amorphous while others were crystalline. Wide-angle X-ray studies have revealed crystalline domains unlike any ever seen in polyethylene materials. Optical studies and material characterizations are underway to determine solution and bulk morphological properties of this family of polymers. If biomaterials can be

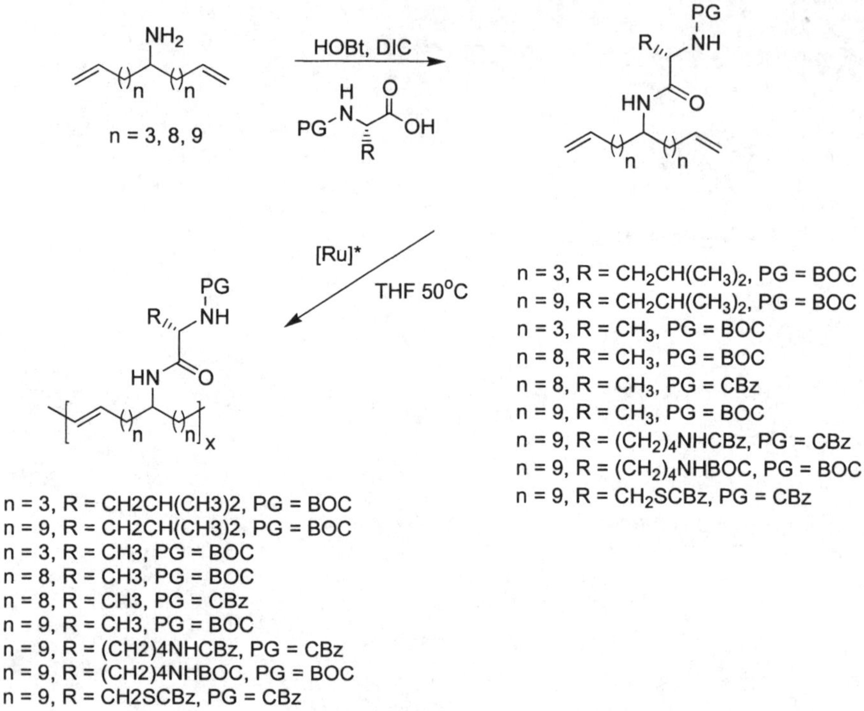

Scheme 15 Amino acid-containing ADMET polymers II (from [170])

produced with selective surface aggregation of amino acid functionality, bioactivity of the bulk polymer could be used in various applications.

5.2
D-*chiro*-Inositol-Based ADMET Polymers

ADMET polymers containing D-*chiro*-inositol have been prepared by polymerizing the acetonide-protected inositol diene followed by deprotection to the D-*chiro*-inositol containing polymer [171]. Synthesis of the monomer precursor was performed using whole-cell fermentation of bromobenzene with *E. coli* JM109.pDTG601A, a recombinant organism expressing toluene dioxogenase [172]. This diol precursor was easily converted to the acetonide protected inositol and then exposed to [Ru] to obtain the protected polymer (Scheme 17). The 18 kg mol^{-1} D-*chiro*-inositol polymer was isolated in good yield upon quantitative deprotection with trifluoroacetic acid, and applications of this polymer in chiral separation and/or catalyst development are being investigated.

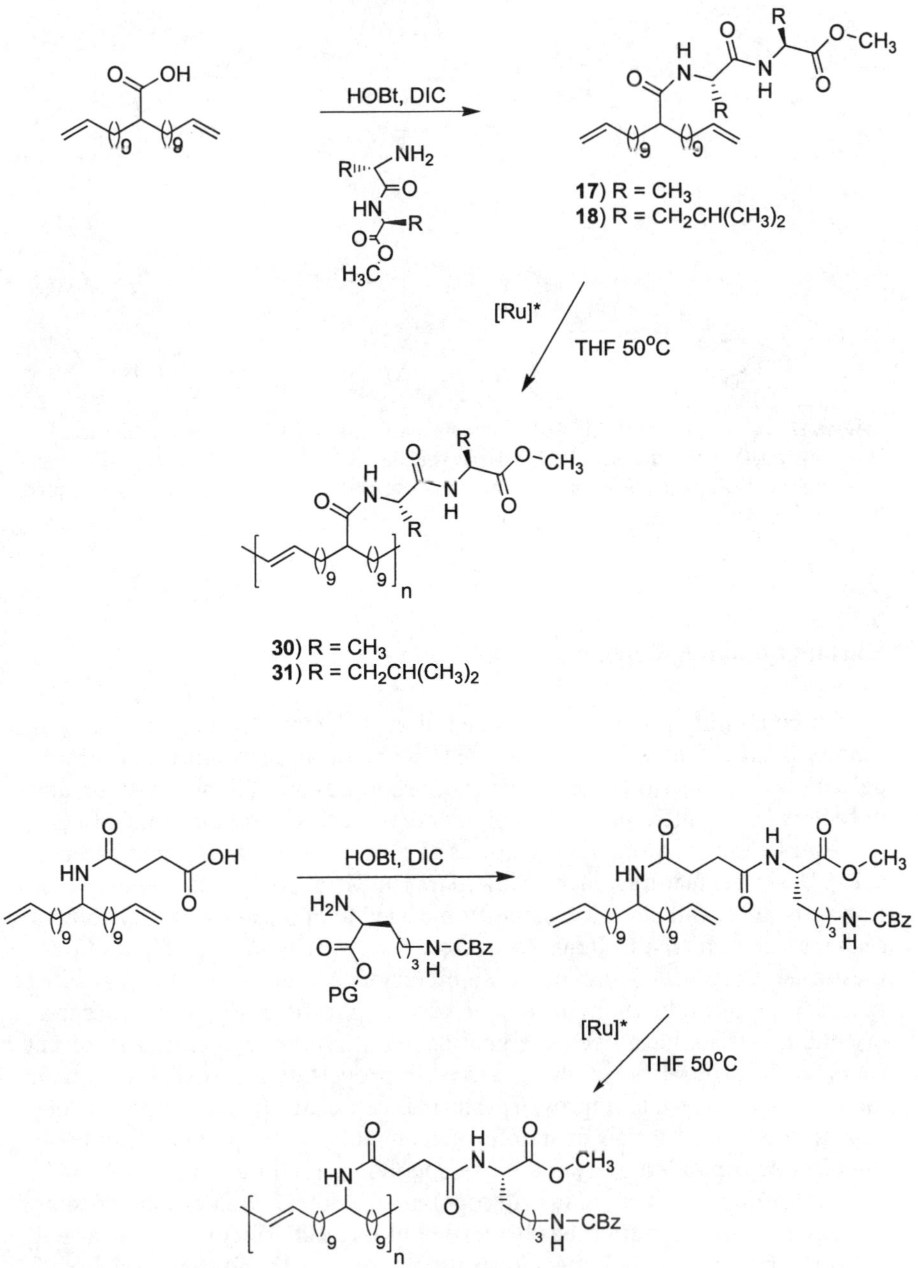

Scheme 16 Peptide-containing ADMET polymers (from [170])

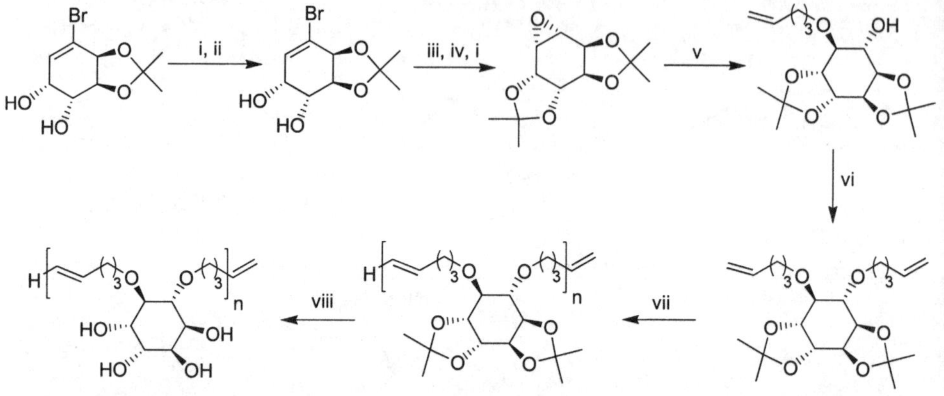

Scheme 17 D-*chiro*-Inositol ADMET polymer synthesis: (i) DMP, pTsOH; (ii) acetone, H_2O, NMO, OsO_4; (iii) Bu_3SnH, AIBN, PhH; (iv) *m*CPBA, CH_2Cl_2; (v) BF_3Et_2O, CH_2Cl_2, 4-penten-1-ol; (vi) DMF, NaH, 5-bromopentene; (vii) [Ru]; (viii) THF-TFA-H_2O 4:1:1 (adapted from [171])

6
Silicon-Containing Polymers

Silicon-containing polymers have been of great interest in the polymer community. Such polymers have found use in biomedical, electronic, ceramic, and agricultural applications, since they offer properties unattainable with organic polymers [173]. Siloxane-based polymers retain flexibility and impact resistance far below operating temperatures of conventional elastomeric materials [174]. Desirable features, such as low glass transition, water repellency, oxidation resistance, and chemical stability, have led to applications as elastomers, sealants, lubricants, and laminates across many areas of industrial and domestic use. Carbosilane polymers exhibit enhanced thermal stability, yielding materials applicable in high-temperature applications. Hybrid inorganic-organic materials have been the goal of many synthetic chemists in recent years, as incorporation of these desirable properties into existing organic materials should lead to improved performance. Relatively few synthetic routes exist to produce materials containing silicon and organic polymers, and with the recently expanded graft and block copolymer methodologies of ADMET polymerization, synthesis of hybrid copolymers in this manner is now feasible.

ADMET polymerization has been used to integrate silicon into linear and network hydrocarbon polymers in an attempt to produce novel materials with enhanced thermal and mechanical stability. While ADMET has been used to produce copolymeric architectures unattainable through conventional methods, application of this polymerization to synthesis is feasible only if the silicon-based functionality does not inhibit metathesis. This research, initiated in the early 1990s by Wagener and colleagues, has shown that the silane and siloxane functionalities can be tolerated by [Mo], and even by the most highly reactive tungsten-based metathesis catalysts, to produce hybrid polymers [175–178].

6.1 PDMS-*b*-Polyoctenamer-*b*-PDMS

ADMET telechelic synthesis has been employed to produce novel block copolymers of polyoctenamer and poly(dimethylsiloxane) (PDMS) [108]. In this synthesis, 4-pentenylchlorodimethylsilane was prepared as a chain terminating endgroup for the telechelic. This endcapping unit was polymerized with 1,9-decadiene using either [Mo] or [Ru] to afford the telechelic oligomers (Scheme 18); NMR, GPC, and vapor pressure osmometry (VPO) confirmed the synthesis of telechelics with M_n values between 1–10 kg mol^{-1}. These highly reactive silane-terminated polyoctenamers were then converted to block copolymers by exposure to hydroxy-terminated PDMS polymer. The single hydroxyl group present at one chain end of the PDMS chain quickly reacted with the disilane telechelics forming two Si–O bonds, releasing HCl, and yielding a novel block copolymer represented as PDMS-*b*-polyoctenamer-*b*-PDMS. Again, NMR, GPC, and VPO were used to confirm the block copolymer struc-

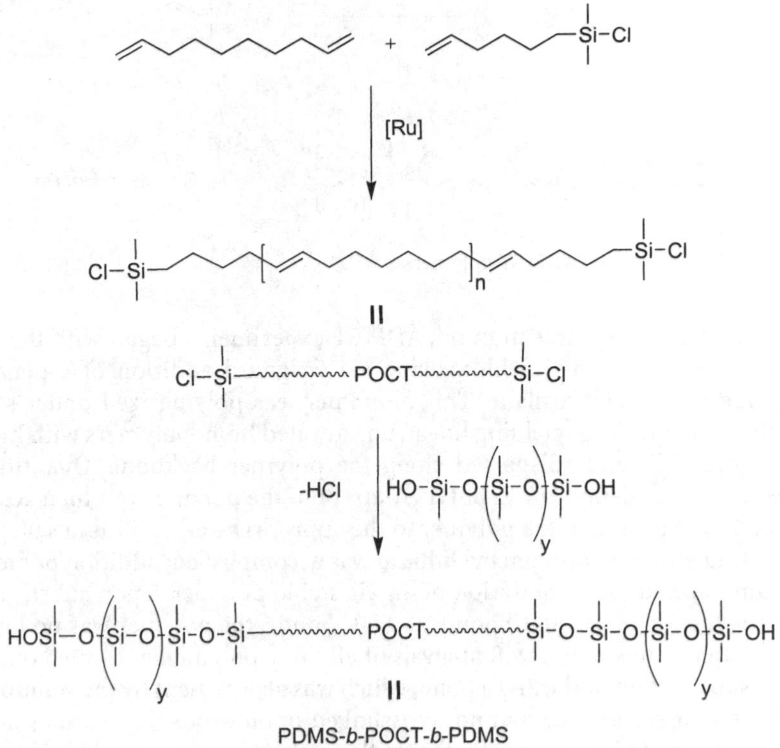

Scheme 18 PDMS-*b*-POCT-*b*-PDMS synthesis

ture; number average molecular weights of around 11.5 kg mol^{-1} were obtained for the block copolymers.

6.2
Carbosilane Polymers via ADMET

Upon synthesis of the telechelic oligomers and further modification of the reactive Si–Cl bond, Wagener and colleagues devised a macromolecular substitution route to polycarbosilanes (Scheme 19) [179]. These experiments involved synthesis of a reactive polymer followed by substitution chemistry throughout the polymer chain. Although many chemical transformations have been attempted on macromolecular systems, few have proven quantitative due to the increased sterics associated with main chain functionality, leading to decreased reactivity and low conversions. However, quantitative substitutions using highly reactive chlorosilanes have been achieved. As an example, West and coworkers demonstrated quantitative substitutions when synthesizing functionalized polycarbosilanes through macromolecular substitution chemistry [180, 181].

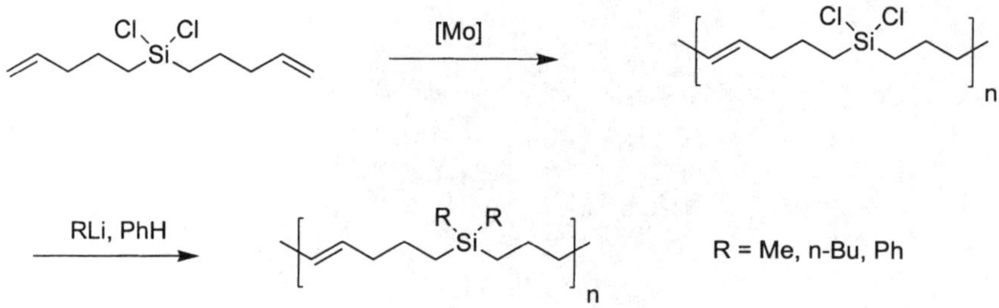

Scheme 19 Carbosilane ADMET polymers

With this past success in mind, ADMET experiments began with the synthesis of di(4-pentenyl)dichlorosilane via Grignard addition of 4-pentenyl bromide to tetrachlorosilane. This monomer was polymerized under stringently dry conditions, yielding linear, unsaturated homopolymers with highly reactive Si–Cl bonds dispersed along the polymer backbone. Quantitative substitution of methyl and n-butyl groups onto the polymer was then accomplished by exposure of the polymer to the appropriate alkyl lithium salt. Only partial substitution with phenyl lithium was accomplished; addition of methyl lithium allowed the preparation of an air stable polymer. Upon substitution and conversion of the Si–Cl bonds to Si–C bonds, the polymer was no longer hydrolytically unstable. NMR analysis of all three polymers indicated quantitative substitution of the Si–Cl bond, which was also evident by the solubilities of the polymers, as they had not crosslinked upon exposure to atmospheric moisture. It had been previously shown that only 1% of unreacted Si–Cl bonds would yield an insoluble network upon exposure to moisture [182]. GPC analyses of the modified parent polymer and all substituted polymers were performed, that indicated number average molecular weights near 20 kg mol^{-1} and polydispersities around 2.0. Production of amorphous hybrid materials via ADMET methodology has proven successful, and this method enables the synthesis of novel functionalized systems through well-known silane substitution chemistry.

6.3
Latent Reactive Carbosilane Polymers

The ADMET monomer di(4-pentenyl)dichlorosilane was also used in the synthesis of carbosilane-based latent reactive polymers [183]. Here, the parent silane was used as a starting point to synthesize a library of silane monomers. Phenoxy, methoxy, ethoxy, and trifluoroethoxy bis chloro silanes were synthesized using standard small substitution techniques, and polymerization was conducted using [Ru]* with a monomer/catalyst ratio of 200:1 (Scheme 20). Only the phenoxy substituted polymer could be characterized via GPC analysis due to the inherent air instabilities of the other alkoxy silane polymers. NMR end group analysis revealed number average molecular weights of 11–15 kg mol^{-1} for the air sensitive polymers, while the phenyl silyl ether polymer was measured at 18 kg mol^{-1} and had a polydispersity of 1.8 via GPC. These polymers illustrate the apparent inactivity of the silicon-oxygen bond during metathesis polymerization.

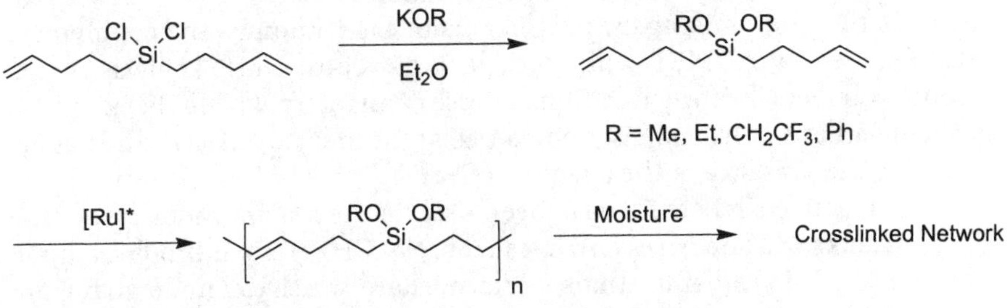

Scheme 20 Latent reactive carbosilane polymers

Homopolymers and copolymers containing carbosiloxane and carbosilane units have been produced that bear latent reactive sites along the chain [184]. Reactive carbosiloxane and unreactive carbosilane homopolymers were first prepared in order to ensure catalyst monomer compatibility and to set end points for copolymer properties. Carbosiloxane homo- and copolymers were synthesized with latent reactivity dispersed throughout the polymer chain in the form of methyl silyl ethers (Scheme 21). It is well known that Si–OMe bonds, although inert during metathesis, can react with atmospheric moisture creating stable Si–O–Si bonds and methanol [185].

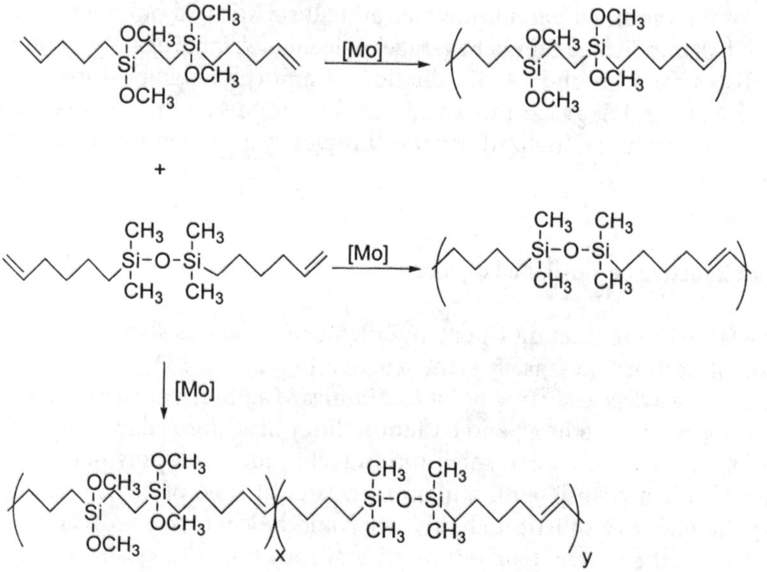

Scheme 21 ADMET carbosilane and carbosiloxane homopolymer and copolymers (from [173])

This type of network synthesis allows for the preparation of a wide variety of copolymers and ultimately networks, offering widely varying chemical and physical properties depending on the comonomer ratio and the density of latent reactive groups along the polymer chain. Linear homo- and copolymers were obtained by ADMET with [Mo] in a glovebox. NMR analysis of the product polymers illustrated the appearance of an internal olefin peak indicative of metathesis polycondensation, as well as the unchanged silyl ether signal indicating the presence of the latent reactive site.

Latent reactive silyl ethers have been used in the sealant industry to cure siloxane caulks and adhesives [185]. A silanol (Si–OH) group is produced upon exposure of the silyl ether to atmospheric moisture, which can undergo further reaction with another silyl ether to create a thermodynamically stable Si–O–Si bond between polymer chains. Varying the molar ratio of the latent reactive carbosilane to the unreactive carbosiloxane allows us to control the crosslink density and material properties of the resultant polymer network. Previous work from this research group has shown the ability of other latent reactive sites to be used to prepare novel hybrid polymers, but the rate of reaction of the Si–Cl bond is much faster than that of the silyl ether, making the latent reactivity hard to control and too fast to allow proper molding of the linear polymer prior to crosslinking [179]. Metathesis polymerization of latent reactive carbosiloxanes into various homo- and copolymeric systems has shown that ADMET is a useful method for synthesizing functional materials. Currently, a family of novel latent reactive monomers is being prepared by copolymerization of carbosilane and carbosiloxane monomers in order to prepare novel hybrid segmented networks for various low temperature applications [186].

7
Conclusion

Recent developments in ADMET polymerization and its use in materials preparation have been presented. Due to the mild nature of the polymerization and the ease of monomer synthesis, ADMET polymers have been incorporated into various materials and functionalized hydrocarbon polymers. Modeling industrial polymers has proven successful, and continues to be applied in order to study polyethylene structure-property relationships. Ethylene copolymers have also been modeled with a wide range of comonomer contents and absolutely no branching. Increased metathesis catalyst activity and functional group tolerance has allowed polymer chemists to incorporate amino acids, peptides, and various chiral materials into metathesis polymers. Silicon incorporation into hydrocarbon-based polymers has been achieved, and work continues toward the application of latent reactive ADMET polymers in low-temperature resistant coatings.

Acknowledgements We would like to acknowledge the NASA Space Grant Consortium and NSF for financial support. We would also like to thank Florence Courchay, Piotr Matloka, and Lani Cardasis for help with the preparation of this manuscript.

References

1. Grubbs RH, Miller SJ, Fu, GC (1995) Acc Chem Res 28:446
2. Grubbs RH, Chang S (1998) Tetrahedron 54:4413
3. Grubbs RH, Trnka TM (2001) Acc Chem Res 34:18
4. Buchmeiser MR (2000) Chem Rev 100:1565
5. Lehman SE, Wagener KB, Grubbs RH (eds)(2003) ADMET polymerization. In: Grubbs RH (ed) Handbook of metathesis, 1st edn, vol 3. Wiley-WCH, Weinheim, p 283
6. Ivin K, Mol JC (1997) Olefin metathesis and metathesis polymerization. Academic, San Diego, CA, Ch 1
7. Eleuterio HS (1991) J Mol Catal 65:55
8. Anderson AW, Merckling, NG (1956) US Patent 2,721,189
9. Anderson AW, Merckling, NG (1956) Chem Abstr 50:3008
10. Calderon N (1967) Tetrahedron Lett 34:3327
11. Calderon N (1972) Acc Chem Res 5:127
12. Fuerstner A (2000) Angew Chem Int Edit 39:3012
13. Wright DL (1999) Curr Org Chem 3:211
14. Schrock RR, Grubbs RH, Feldman J, Cannizzo LF (1987) Macromolecules 20:1169
15. Schrock RR (1988) Polym Mater Sci Eng 58:92
16. Wagener KB, Boncella JM, Nel JG, Konzelman J (1990) Macromolecules 23:5155
17. Wagener KB, Boncella JM, Nel JG (1991) Macromolecules 24:2649

18. Wagener KB, Puts RD, Smith DW Jr (1991) Makromol Chem Rapid Comm 12:419
19. Wagener KB, Smith DW (1991) Macromolecules 24:6073
20. Wagener KB, Smith DW (1993)Macromolecules 26:1633
21. Chauvin Y, Herisson JL (1971) Makromol Chem 141:161
22. Katz TJ, McGinnis J (1975) J Am Chem Soc 97:1592
23. Kress J, Osborn JA, Wesolek M (1982) J Chem Soc Chem Commun 514
24. Kress J, Osborn JA (1987) J Am Chem Soc 109:3953
25. Schrock RR, Murdzek JS, Bazan GC, Robbins J, DiMare M, O'Regan M (1990) J Am Chem Soc 112:3875
26. Grubbs RH, Nguyen ST (1993) J Am Chem Soc 115:9858
27. Arduengo III AJ (1999) Acc Chem Res 32:913
28. Scholl M, Trnka TM, Morgan JP, Grubbs RH (1999) Tetrahedron Lett 40:2247
29. Weskamp T, Schattenmann WC, Spiegler M, Herrman WA (1998) Angew Chem 110:2631
30. Herrmann WA, Kocher C (1997) Angew Chem 109:2256
31. Weskamp T, Kohl FJ, Herrmann WA (1999) J Organomet Chem 582:362
32. Weskamp T, Kohl FJ, Hieringer W, Gleich D, Herrmann WA (1999) Angew Chem 111: 2573
33. Frenzel U, Weskamp T, Kohl FJ, Schattenmann WC, Nuyken O, Herrmann WA (1999) J Organomet Chem 586:263
34. Hamilton JG, Frenzel U, Kohl FJ, Weskamp T, Rooney JJ, Hermann WA, Nuyken O (2000) J Organomet Chem 606:8
35. Baratta W, Herdtweck E, Herrmann WA, Rigo P, Schwartz J (2002) Organometallics 21:2101
36. Scholl M, Lee CW, Grubbs RH (1999) Org Lett 1:953
37. Lehman Jr SE, Schwendeman JE, O'Donnell PM, Wagener KB (2003) Inorg Chim Acta 345:190
38. Lehman Jr SE, Schwendeman JE, O'Donnell PM, Wagener KB (2003) Polym Prepr 44:947
39. Smith JA, Brzezinska KR, Valenti DJ, Wagener KB (2000) Macromolecules 33:3781
40. Hopkins TE, Wagener KB (2004) Macromolecules 37:1180
41. Anhaus JT, Clegg W, Collingwood SP, Gibson VC (1991) J Chem Soc Chem Commun 1720
42. Watson MD, Wagener KB (1999) J Polym Sci Part A 37:1857
43. Watson MD, Wagener KB (2000) Macromolecules 33:1494
44. Boffa LS, Novak BM (2000) Chem Rev 100:1479
45. Odian G (1991) Principles of polymerization, 3rd edn. Wiley, New York, Ch 6
46. Tempel DJ, Johnson LK, Huff RL, White PS, Brookhart MJ (2000) J Am Chem Soc 122: 6686
47. Sworen JC, Smith JA, Wagener KB, Baugh LS, Rucker SP (2003) J Am Chem Soc 125:2228
48. Wunderlich B, Poland DJ (1963) J Polym Sci Part A 1:357
49. Alamo RG, Maldelkern L (1989) Macromolecules 22:1273
50. Alamo RG, Maldelkern L, Stack GM, Kronke C, Wegner G (1994) Macromolecules 27:147
51. Mirabella Jr. FM, Ford EA (1987) J Polym Sci Part B 25:777
52. Ke B (1962) J Polym Sci 61:47
53. Ke B (1960) J. Polym Sci 42:15
54. Kawaguchi T, Ito T, Kawai H, Keedy D, Stein RS (1968) Macromolecules 1:126
55. Schumacher M, Lovinger AJ, Agarwal P, Wittman JC, Lotz B (1994) Macromolecules 27:6956
56. Hashimot T, Prud'homme RE, Stein RS (1973) J Polym Sci Pol Phys 11:709
57. Fawcett EW, Gibson RQ, Perrin MH, Patton JG, Williams EG (1937) Brit Pat 2,816,883
58. Mirabella FM (2001) J Polym Sci Pol Phys 39:2800

59. Kravchenko R, Waymouth RM (1998) Macromolecules 31:1
60. Jungling S, Koltzenburg S, Mulhaupt RJ (1997) J Polym Sci Pol Chem 35:1
61. Rix F, Brookhart M (1995) J Am Chem Soc117:1137
62. Schmidt GF, Brookhart M (1985) J Am Chem Soc 107:1443
63. Brookhart M, Volpe Jr AF, Lincoln DM, Horvath IT, Millar JM (1990) J Am Chem Soc 112:5634
64. Kim JS, Pawlow JH, Wojcinski II LM, Murtuza S, Kracker S, Sen AJ (1998) J Am Chem Soc 120:4049
65. Gates DP, Svejda SA, Onate E, Killian CM, Johnson LK, White PS, Brookhart M (2000) Macromolecules 33:2320
66. Mattice WL (1983) Macromolecules 16:487
67. Roedal MJ (1953) J Am Chem Soc 75:6110
68. O'Gara JE, Wagener KB, Hahn SF (1993) Polym Prep 34:406
69. O'Gara JE, Wagener KB, Hahn SF (1993) Makromol Chem Rapid Commun 14:657
70. Valenti DJ, Wagener KB (1997) Macromolecules 30:6688
71. Smith JA, Brzezinska KR, Valenti DJ, Wagener KB (2000) Macromolecules 33:3781
72. Sworen JC, Smith JA, Wagener KB, Baugh LS, Rucker SP (2003) J Am Chem Soc 125:2228
73. Odian G (1991) Principles of polymerization, 3rd edn. Wiley, New York, Ch 4, p 40
74. Zutty NL, FaucherJA, Bonotto S, Mark HF (eds)(1967) Encyclopedia of polymer science and technology. Wiley, New York
75. Odian G (1991) Principles of polymerization, 3rd edn. Wiley, New York, p 460
76. Chowdhury F, Haigh JA, Maldelkern L, Alamo RG (1998) Polym Bull 41:463
77. Bistac S, Kunemann P, Schultz J (1998) Polymer 39:4875
78. Smith GD, Liu F, Devereaux RW, Boyd RH (1992) Macromolecules 22:2699
79. Watson MD, Wagener KB (2000) Macromolecules 33:5411
80. Hilmyer MA, Laredo WR, Grubbs RH (1995) Macromolecules 28:6311
81. Watson MD, Wagener KB (2000) Macromolecules 33:8963
82. Watson MD, Wagener KB (2000) Macromolecules 33:3196
83. Wagener KB, Valenti DJ, Hahn SF (1997) Macromolecules 30:6688
84. Younkin TR, Connor EF, Henderson JI, Friedrich SK, Grubbs RH, Bansleben DA (2000) Science 287:460
85. Mani R, Burns CM (1991) Macromolecules 24:5476
86. Venditto V, DeTullio G, Izzo L, Olivia L (1998) Macromolecules 31:4027
87. D'Aniello C, Decandia F, Olivia L Vittoria V (1995) J Appl Polym Sci 58:1701
88. Sernetz FG, Mulhaupt R (1997) Macromolecules 30:1562
89. Zutty NL, FaucherJA, Bonotto S, Mark HF (eds)(1967) Encyclopedia of polymer science and technology. Wiley, New York, p 386
90. Bowmer TN, Tonelli AE (1985) Polymer 26:1195
91. Pourahmady N, Bak PI (1992) J Macromol Sci Pure Appl Chem A29:959
92. Wegner G, Gutzler F (1980) Colloid Polym Sci 258:776
93. Gomez MA, Tonelli AE, Lovingerr AJ, Schilling FC, Cozine MH, Davis DD (1989) Macromolecules 22:4441
94. Shinoda H, Miller PJ, Matyjaszewski K (2001) Macromolecules 34:3186
95. Zhu L, Cheng SZD, Calhoun BH, Ge Q, Quirk RP, Thomas EL, Hsiao BS, Yeh F, Lotz B (2000) J Am Chem Soc 122:5957
96. Matyjaszewski K, Teodorescu M, Miller PJ, Peterson ML (2000) J Polym Sci Pol Chem 38:2440
97. Patten TE, Matyjaszewski K (1998) Adv Mater 10:901
98. Webster OW (1991) Science 251:887
99. Shinoda H, Matyjaszewski K (2001) Macromolecules 34:6243

100. Schultz GO, Milkovich RJ (1982) J Appl Polym Sci 27:4773
101. Jerome R (1999) Macromol Chem Phys 200:156
102. O'Donnell PM, Brzezinska K, Powell D, Wagener KB (2001) Macromolecules 34:6845
103. O'Donnell PM, Wagener KB (2003) J Polym Sci Pol Chem 41:2816
104. Bielawski CW, Morita T, Grubbs RH (2000) Macromolecules 33:678
105. Bielawski CW, Louie J, Grubbs RH (2000) J Am Chem Soc 122:12872
106. Goethals EJ (1989) Telechelic polymers: synthesis and characterization. CRC, Boca Raton, FL
107. Schwendeman JE, Wagener KB (2002) Polym Prepr 43:280
108. Brzezinska KR, Wagener KB, Burns GT (1999) J Polym Sci Pol Chem 37:849
109. Schwendeman JE (2002) Amorphous telechelic hydrocarbon diols and ethylene-based model copolymers via acyclic diene metathesis. PhD Dissertation, University of Florida, Gainesville, FL
110. Wagener KB, Nel JG, Duttweiler RP, Hilmyer MA, Boncella JM, Konzelman J, Smith DW Jr, Puts R, Willoughby L (1991) Rubber Chem Technol 64:83
111. Hillmayer MA, Nguyen ST, Grubbs RH (1997) Macromolecules 30:718
112. Brzezinska KR, Deming TJ (2001) Macromolecules 34:4348
113. Billot JP, Douy A, Gallot B (1977) Makromol Chem 178:1641
114. Morita T, Maughon BR, Bielawski CW, Grubbs RH (2000) Macromolecules 33:6621
115. Deming TJ (2000) J Polym Sci Pol Chem 38:3011
116. Curtin SA, Deming TJ (1999) J Am Chem Soc 121:7427
117. Deming TJ (1997) Nature 390:386
118. Yu YS, Jerome R, Fayt R, Teyssie P (1994) Macromolecules 27:5957
119. Tamura H, Nakayama A (2002) J Macromol Sci Pure App Chem 39:745
120. Choi TL, Rutenberg IM, Grubss RH (2002) Angew Chem 114:3995
121. Choi TL, Rutenberg IM, Grubss RH (2002) Angew Chem Int Edit 41:3839
122. Demel S, Slugovc C, Stelzer F, Fodor-Csorba K, Galli G (2003) Macromol Rapid Commun 24:636
123. Allcock HR (1996) Macromolecular design of polymeric materials. Marcel Dekker, New York, p 447
124. Allcock HR (1998) Appl Organomet Chem 12:659
125. Allcock HR (1972) Phosphorus-nitrogen compounds. Cylic, linear, and polymeric systems. Academic, New York.
126. Kumar D, Khullar M, Gupta AD (1993) Polymer 34:3025
127. Kumar D, Gupta AD (1995) Macromolecules 28:6323
128. Chen-Yang YW, Chuang YH (1993) Phosphorus Sulfur 76:261
129. Miyata K, Muraoka K, Itaya T, Tanigaki T, Inoue K (1996) Eur Polym J 32:1257
130. Dez I, de Jagger R (1997) Phosphorus Sulfur 130:1
131. Tunca U, Hizal G (1998) J Polym Sci Pol Chem 36:1227
132. Allcock HR, Lampe FW, Mark JE (2003) Contemporary polymer chemistry. Pearson Education, Upper Saddle River, NJ, p 244
133. Allcock HR, Laredo WR, deDenus CR, Taylor JP (1999) Macromolecules 32:7719
134. Allcock HR, Kellam EC, Hofmann MA (2000) Polym Prepr 41:1233
135. Allcock HR, Kellam EC, Hofmann MA (2001) Macromolecules 34:5140
136. Allcock HR, Kellam EC (2002) Macromolecules 35:40
137. Allcock HR, Lampe FW, Mark JE (2003) Contemporary polymer chemistry. Pearson Education, Upper Saddle River, NJ, p 735
138. Malls JJM, Walsh CA, Greenham NC, Marseglia EA, Friend RH, Moratti SC, Holmes AB (1995) Nature 376:498
139. Granstrom M, Berggren M, Ignanas O (1995) Science 267:1479

140. Burns PL, Holmes AB, Kraft A, Bradley DDC, Brown AR, Friend RH, Gymer RW (1992) Nature 356:47
141. Burroughes JH, Bradley DDC, Brown AR, Marks RN, Mackay K, Friend RH, Burns PL, Holmes AB (1990) Nature 347:539
142. Kaino T, Kurihare T, Saito S, Tsutsui T, Tokito S (1989) Phys Lett 54:1619
143. Karasz FE, Wnek GE, Chien JC, Liilya CP (1979) Polymer 20:1441
144. Peetz R, Strachota A, Thorn-Csanyi E (2003) Macromol Chem Phys 204:1439
145. Thorn-Csanyi E, Herzog O (2004) J Mol Catal A 213:123
146. Birchall AC, Bush SM, North M (2001) Polymer 42:375
147. Langer R (2000) Acc Chem Res 33:94
148. Okamoto Y, Nakano T (1995) Chem Rev 94:349
149. Pino P, Lorenzi GP (1960) J Am Chem Soc 82:4745
150. Natta G, Pino P, Mazzanti G, Corradini P, Giannini U (1955) 42:712
151. Satoh T, Yokota K, Kakuchi T (1995) Macromolecules 28:4762
152. Kakuchi T, Morimoto Y, Uesaka T, Yokota K (1995) Macromolecules 28:6378
153. Ito Y, Ihara E, Murakami M (1992) Angew Chem Int Edit 31:1509
154. Yokota K, Haba O, Satoh T (1995) Macromol Chem Phys 196:2383
155. Kakuchi T, Harada Y, Hashimoto H, Satoh T, Yokota K (1994) J Macromol Sci Pure Appl Chem A31:751
156. Schrock RR, Hoveyda AH (2003) Angew Chem Int Edit 42:4592
157. La DS, Sattely ES, Ford JG, Schrock RR, Hoveyda AH (2001) J Am Chem Soc 123:7767
158. Weatherhead GS, Ford JG, Alexanian EJ, Schrock RR, Hoveyda AH (2000) J Am Chem Soc 122:1828
159. Buchmeiser M, (2003) Well-defined transition metal catalyst for metathesis polymerization. In: Rieger B, Baugh LS, Kacker S, Striegler S (eds) Late transition metal polymerization catalysis. Wiley-VCH, Weinheim, p 164
160. Sanda F, Endo T (1999) Macromol Chem Phys 200:2651
161. Morcellet-Sauvage J, Morcellet M, Loucheux C (1981) Makromol Chem 182:949
162. Morcellet-Sauvage J, Morcellet M, Loucheux C (1982) Makromol Chem 183:821
163. Bottiglione V, Morcellet M, Loucheux C (1980) Makromol Chem 181:485
164. Lekchiri A, Morcellet J, Morcellet M (1987) Macromolecules 20:49
165. Methenitis C, Morcellet J, Morcellet M, Pnuematikakis G (1994) Macromolecules 27:1455
166. Green MM, Garetz BA (1984) Tetrahedron Lett 25:2831
167. Maynard HD, Okada SY, Grubbs RH (2001) J Am Chem Soc 123:1275
168. Maynard HD, Okada SY, Grubbs RH (2000) Macromolecules 33:6239
169. Hopkins TE, Pawlow JH, Koren DL, Deters KS, Solivan SM, Davis JA, Gomez FJ, Wagener KB (2001) Macromolecules 34:7920
170. Hopkins TE, Wagener KB (2004) Macromolecules 37:1180
171. Hudlicky T, Bui VP (2004) Tetrahedron 60:641
172. Zylstra GJ, Gibson DT (1989) J Biol Chem 264:14940
173. Allcock HR (1989) Comprehensive polymer science, 1st edn, vol 4. Pergamon, New York, p 8
174. Odian G (1991) Principles of polymerization. Wiley, New York, p 138
175. Smith DW Jr, Wagener KB (1993) Macromolecules 26:1633
176. Smith DW Jr, Wagener KB (1993) Macromolecules 26:3533
177. Smith Jr DW (1992) Unsaturated organosilicon polymers via acyclic diene metathesis (ADMET) condensation polymerization. PhD Dissertation, University of Florida, Gainesville, FL
178. Smith DW Jr, Wagener KB (1991) Macromolecules 24:6073

179. Church AC, Pawlow JH, Wagener KB (2002) Macromolecules 35:5746
180. Koe JR, Montonaga M, Fujiki M, West R (2001) Macromolecules 34:706
181. Herzog U, West R (1999) Macromolecules 32:2210
182. Caprino JC, Macander RF Morton M (eds)(1999) Rubber technology, 3rd edn. Kluwer Academic, Boston, MA
183. Church AC, Pawlow JH, Wagener KB (2003) Macromol Chem Phys 204:32
184. Brzezinska KR, Schitter R, Wagener KB (2000) J Polym Sci Pol Chem 38:1544
185. Rochow EG (1989) The chemistry of silicon, 1st edn, vol 4. Pergamon, New York, p 10
186. Matloka P (2004) The chemistry of latent reactive polycarbosilane/polycarbosiloxane elastomers via acyclic diene metathesis (ADMET) polymerization. Masters Dissertation, University of Florida, Gainesville, FL

Received: April 2004

Liquid Crystalline Polymers by Metathesis Polymerization

Gregor Trimmel (✉) · Silvia Riegler · Gernot Fuchs · Christian Slugovc · Franz Stelzer

Institute for Chemistry and Technology of Organic Materials (ICTOS), Graz University of Technology, Stremayrgasse 16, 8010 Graz, Austria
trimmel@ictos.tu-graz.ac.at, franz.stelzer@tugraz.at

1	Introduction	44
1.1	Liquid Crystalline Polymers	44
1.2	Initiator Systems	46
2	Side chain liquid crystalline polymers (SCLCPs)	47
2.1	Biphenyl-Based Mesogens	48
2.1.1	Influence of Spacer Length	49
2.1.2	Influence of Mesogen Density	52
2.1.3	Influence of Tacticity	54
2.1.4	Magnetic Orientation of SCLCPs	56
2.1.5	LC Elastomers	58
2.1.6	SCLCPs with Different Backbones	58
2.1.7	Comparison of Cyanobiphenyl-Based SCLCPs Prepared by Different Methods	59
2.1.8	SCLCP Block Copolymers	61
2.2	Terminally Attached Mesogens of Various Kinds	63
2.2.1	Block Copolymers	65
2.3	Laterally-Attached Mesogens	67
2.3.1	Block Copolymers	75
2.4	Discotic Mesogens	77
2.5	Dendritic Side Chains	78
3	Main Chain Liquid Crystalline Polymers by ADMET und ALTMET	81
4	Conclusion and Outlook	84
References		85

Abstract Liquid crystalline polymers (LCPs) have gained attention as materials with interesting optical, mechanical and rheological properties. This review summarizes research on thermotropic liquid crystalline polymers synthesized by metathesis polymerization techniques. Ring opening metathesis polymerization (ROMP) is a versatile tool for preparing side chain liquid crystalline polymers (SCLCPs) with different mesogenic units. The living character of ROMP allows us to design well-defined polymers and copolymers with controlled molecular weight and molecular weight distributions. The mesophase type and the thermal transition temperatures are influenced not only by the attached mesogenic moiety but also by the stiffness of the polymer backbone, the spacer lengths between the main chain and the liquid crystalline entity, the Z/E isomerism, the tacticity, and so on. In addition, the use of acyclic diene metathesis polymerization (ADMET) or alternating diene metathesis polycondensation (ALTMET) offer new possibilities for creating new main chain liquid crystalline polymers (MCLCPs). A short discussion of new trends and potential applications for these polymers concludes the review.

Keywords Liquid crystalline polymers · Metathesis · ROMP · ADMET · ALTMET

List of Abbreviations
LC	Liquid Crystal(line)
ROMP	Ring-opening metathesis polymerization
ADMET	Acyclic diene metathesis polymerization
ALTMET	Alternating diene metathesis polycondensation
MCLCP	Main chain liquid crystalline polymer
SCLCP	Side chain liquid crystalline polymer
mru	molecular repeating unit
k	crystalline
g	glassy
s	smectic
n	nematic
i	isotropic
D_h	discotic hexagonal
Φ_H	hexagonal columnar
PDI	polydispersity index (M_w/M_n)
DP	degree of polymerization
T_g	glass transition temperature
DSC	differential scanning calorimetry
SEC	size exclusion chromatography
POM	polarized optical microscopy
WAXD	wide angle X-ray diffraction
SAXS	small angle X-ray scattering

1
Introduction

1.1
Liquid Crystalline Polymers

Since their discovery in the nineteenth century [1], liquid crystals have fascinated scientists due to their unusual properties and their wide range of potential applications, especially in optoelectronics. LC systems can be divided into two categories: thermotropic LC phases and lyotropic LC phases. Thermotropic LC systems result from anisotropic molecules or molecular parts (so called mesogens or mesogenic moieties, respectively), and form "liquid crystalline" phases between the solid state and the isotropic liquid state, where they flow like liquids but possess some of the characteristic physical properties of crystals [2]. One can distinguish between rod-like (calamitic) and disk-like (discotic) mesogens, which are the two most important mesogen classes.

Lyotropic liquid crystal phases are formed by amphiphilic molecules (surfactants, block copolymers) in solution, driven by repulsive forces between hydrophobic and hydrophilic parts. In a polar solvent, the hydrophilic parts associate with the solvent, whereas the hydrophobic parts interact to form the interiors of micelles (as in low-molecular surfactants). Micelles can be spherical, rod-like or discotic in shape. The contour of the micelle is determined by the relative sizes of the hydrophilic and hydrophobic groups. Micelle shapes are influenced by solvent, concentration and temperature.

Liquid crystallinity can be attained in polymers of various polymer architectures, allowing the chemist to combine properties of macromolecules with the anisotropic properties of LC-phases. Mesogenic units can be introduced into a polymer chain in different ways, as outlined in Fig. 1. For thermotropic LC systems, the LC-active units can be connected directly to each other in a condensation-type polymer to form the main chain ("main chain liquid crystalline polymers", MCLCPs) or they can be attached to the main chain as side chains ("side chain liquid crystalline polymers", SCLCPs). Calamitic (rod-like) as well as discotic mesogens have successfully been incorporated into polymers. Lyotropic LC-systems can also be formed by macromolecules. Amphiphilic block copolymers show this behavior when they have well-defined block structures with narrow molecular weight distributions.

Liquid crystalline polymers (LCPs) have gained attraction as materials with interesting optical, mechanical and rheological properties [3–7]. This review summarizes research on thermotropic liquid crystalline polymers synthesized by metathesis routes, as this chemistry has proven to be a versatile way to build up well-defined polymer architectures [8]. Recent results promise to expand the possible uses of these methods.

Ring opening metathesis polymerization (ROMP) can be used to build up SCLCPs using various mesogenic units. ROMP-derived SCLCPs exhibit a num-

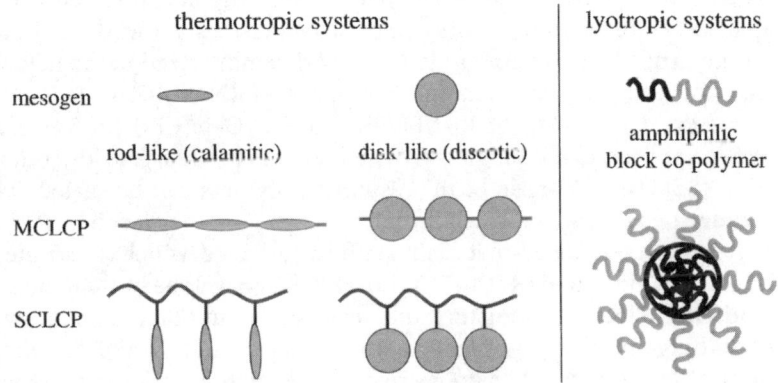

Fig. 1 Possible architectures for thermotropic and lyotropic LCPs

ber of additional structural features compared to polymers prepared by other polymerization methods: Z/E-isomerism, head/tail connections, tacticity and the substitution pattern of the monomers, to name but a few. Some exceptional properties result from special variations of these structural elements. Statistical copolymers and (due to the fact that ROMP is a living polymerization method) block copolymers with narrow molecular weight distributions can easily be prepared thus combining the properties of the co-monomers. Depending on the initiating system, almost all parameters can be modified in a well-defined way. In addition, the use of ADMET or ALTMET offers new possibilities for creating MCLCPs that are not accessible by other methods.

1.2
Initiator Systems

Olefin metathesis is one of the few fundamentally new organic reactions discovered over the last few decades that has revolutionized organic and polymer chemistry. Olefin metathesis provides a convenient and reliable way to synthesize unsaturated molecules that are often hard to prepare by any other method. A number of reviews [8–17] and books [18–20] have been published in this area, all of which focus on the ever-increasing use of olefin metathesis in organic synthesis and polymer chemistry. Particularly in the latter research area, ROMP has become a powerful and popular method to synthesize polymers with narrow molecular weight distributions. Due to the living nature of polymerizations initiated by state-of-the-art initiators, well-defined diblock, triblock or multiblock copolymers are available today.

It is not the aim of this review article to discuss the chemistry of the initiators (or catalysts) for alkene metathesis, but some comments on the initiators used might contribute to our understanding of why this remarkable polymerization technique has not attracted more attention up to now.

Initiators can be subdivided into ill-defined and well-defined initatorsystems. The latter ones dominate in this review. Ill-defined initiators like $WCl_4(OAr)_2$-$PbEt_4$ (Ar=2,6-diphenylphenyl) have only been used in isolated examples [21]. Preferred initiators comprise two classes: molybdenum-based systems and ruthenium compounds. Most work summarized in the following sections is based on molybdenum alkylidenes $Mo(NAr')(CHCMe_2R)(OR')_2$ (Ar'=2,6-i-PrC_6H_3; R=Me, Ph; R'=t-But, $CMe_2(CF_3)$, $CMe(CF_3)_2$) **1–5** (Fig. 2).

These complexes exhibit high activity and, in most cases, provide complete initiation [22]. The E/Z-ratio of the resulting polymers can be varied over a broad range depending on which initiator is used. Another fact is that cyano groups, often present in LC-materials, are fully tolerated, which constitutes an advantage over initiator **6** ($RuCl_2(PCy_3)_2(CHPh)$, Cy=cyclohexyl). On the other hand, initiator **6** tolerates moisture and air to an extent that allows its convenient handling and the use of less rigorously dried solvents and monomers. Polymerizations with **6** are characterized by moderate reaction rates. The introduction of $RuCl_2(H_2IMes)(PCy_3)(CHPh)$ (H_2IMes=N,N-di(mesityl)-4,5-di-

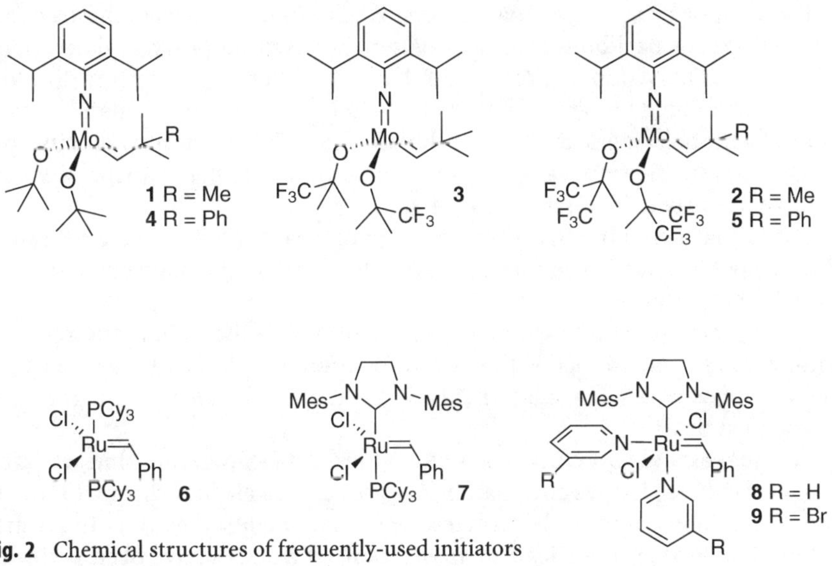

Fig. 2 Chemical structures of frequently-used initiators

hydroimidazolin-2-ylidene) **7** (Fig. 2) has improved not only the activity but also the functional group tolerance of ruthenium based initiators – CN groups are unrestrictedly tolerated by this catalyst [23, 24]. Interestingly, **7** is sparely used for the preparation of liquid crystalline polymers, which might be due to the relatively low initiation efficiency of **7** in ROMP reactions. Very recent developments in the field of initiators will certainly provide a new impetus in metathesis polymerization. Pyridine-based ruthenium benzylidenes like $RuCl_2(H_2IMes)(pyridine)_2(CHPh)$ (**8**) [25] or $RuCl_2(H_2IMes)(3$-bromo-pyridine$)_2(CHPh)$ (**9**) [26-28] provide complete initiation, high activity, and extraordinary functional group tolerance. This new generation of initiators will allow the reliable synthesis of an even wider array of liquid crystalline polymers of different architectures without important restrictions on the choice of monomer.

2
Side chain liquid crystalline polymers (SCLCPs)

SCLCPs combine liquid crystalline properties and polymeric behavior in one material. If the mesogenic unit is fixed directly to the polymer main chain, the motion of the liquid crystalline side chain is coupled with the motion of the polymer backbone, preventing the formation of a LC mesophase. Therefore, Finkelmann and Ringsdorf proposed that the introduction of a flexible spacer between the main chain and the mesogenic unit would decouple their motions, allowing the mesogenic moiety to build up an orientational order [29, 30].

Based on this concept, a broad variety of SCLCPs have been synthesized that utilize different backbones, such as polymers based on poly(acrylate), poly(siloxane) or poly(phosphazene) and that have different rod-like or disc-like mesogenic units. The type of liquid crystalline phase formed is mainly determined by the mesogenic unit chosen, but the main chain (rigidity, tacticity), the spacer (rigidity, length), and the molecular weight and polydispersity also strongly influence the resulting LC phase [31].

The liquid crystalline state only exists above the glass transition temperature T_g of the polymer, and it competes with the tendency of the main chain to form a random orientation.

Probably the most interesting feature of SCLCPs is their ability to freeze an anisotropic alignment below the glass transition, coupled with the fluidity of the mesophase [32]. This alignment can be attained by electric, magnetic or mechanical fields.

This makes these polymers useful for optical data storage, non-linear optics, stationary phases for gas chromatography, supercritical-fluid chromatography and HPLC, solid polymer electrolytes, separation membranes and display materials. Syntheses, properties and applications of SCLCPs have been reviewed [5, 6, 31, 33, 34]. Different polymerization techniques and build-up strategies have been used to synthesize these polymers. To synthesize LC-polymers with well-defined architectures, living polymerization techniques such as anionic and group transfer polymerization, cationic polymerization, ring-opening metathesis polymerization or controlled radical polymerization techniques are necessary. Furthermore, these techniques allow us to couple SCLCP blocks with other blocks and to introduce additional properties or functionalities into the material. Using this approach, lyotropic and thermotropic LC properties can be combined into one polymer [31, 35, 36]. This kind of polymer is interesting because of the appearance of both ordering principles, microphase separation and thermotropic LC-mesophases in the same polymer.

Ring opening metathesis polymerization (ROMP) has proven to be a convenient technique that allows us to synthesize a wide variety of SCLCPs [8].

2.1
Biphenyl-Based Mesogens

The liquid crystalline phases of a SCLCP are influenced by the design of the polymer: the polymer backbone, the spacer, the linking or anchoring group (A), the mesogen (M) and the number of mesogens per monomer unit ("mesogen density"); see Fig. 3. In addition, the mesophase depends on the molecular weight, the molecular weight distribution, the Z/E ratio and the tacticity of the polymer. Because of the numerous articles published on SCLCPs with biphenyl units as mesogens (M=C_6H_4-C_6H_4-R, R=CN, OCH$_3$), we will now discuss these features in more detail.

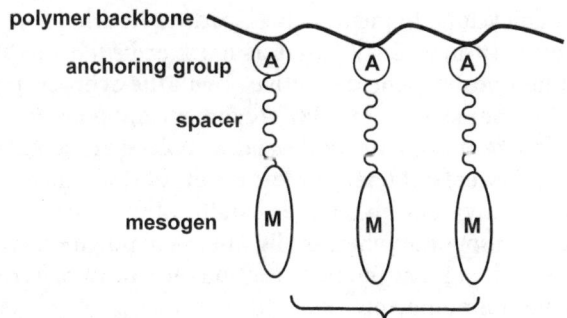

Fig. 3 Schematic design of a SCLCP

2.1.1
Influence of Spacer Length

Biphenyl mesogens with R=CN [(4′-cyanobiphenyl-4-yl-)oxy-] or R=OCH$_3$ [(4′-methoxybiphenyl-4-yl)-oxy-] were attached to norborn-2-ene moieties, and separated by different spacer lengths and carboxylic acid ester, ether or imido-groups as anchoring moieties, as shown in Fig. 4.

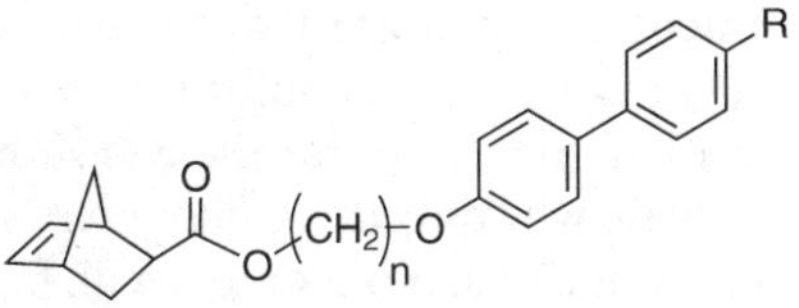

I-n: R = OCH$_3$ n = 2-12
II-n: R = CN n = 3-7, 9-12

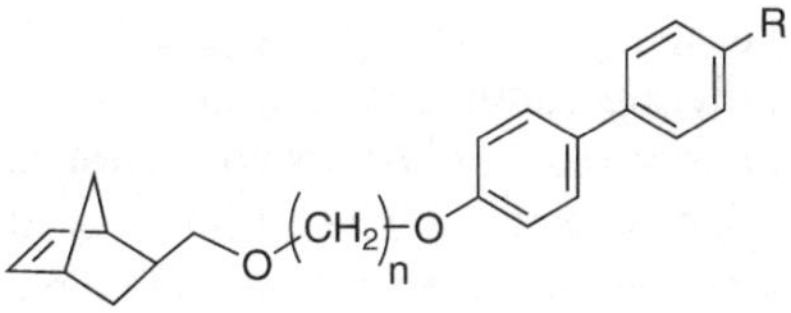

III-n: R = OCH$_3$ n = 4-6

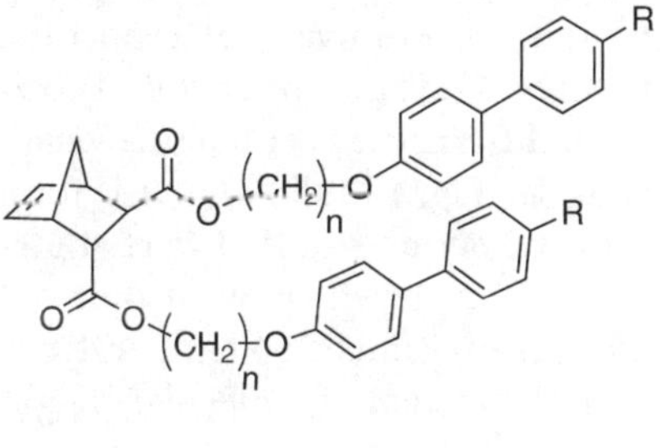

IV-n: R = CN n = 2-12
V-n: R = OCH$_3$ n = 4-10

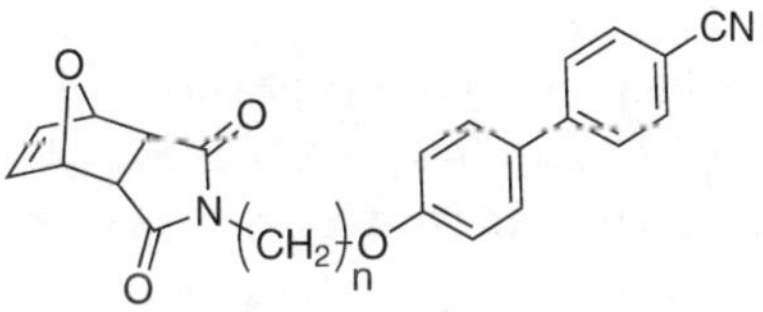

VI-n: n = 2-8

Fig. 4 Chemical structures of monomers with biphenyl mesogens

The ROMP of bicyclo[2.2.1]hept-5-ene-2-carboxylic acid (4'-cyanobiphenyl-4-yl)oxyalkyl ester (**I-n**, *n*=2–12) used as a 3:1 *exo/endo* mixture (Fig. 4) – a monosubstituted norbornene derivative – with the Schrock-type initiator **1** was described by Komiya et al. [37, 38]. Oligomers and polymers with degrees of polymerization from 5 to 100 and narrow molecular weight distributions (M_w/M_n=1.05–1.28) were obtained (for the results of DSC and SEC see Table 1). All oligomers and polymers with spacer lengths of up to *n*=10 show an enantiotropic nematic mesophase, whereas oligomers and polymers with *n*=11 or 12 have an enantiotropic smectic phase. Exceptions are the two shortest oligomers of **I-12**, that exhibit a monotropic smectic mesophase. Polymers with an odd number of methylene units in the spacer display higher isotropization temperatures than those with an even number. This odd-even effect vanishes when the spacer length is greater than *n*=6. Glass transition temperatures decrease whereas isotropization temperatures increase with increasing molecular weight, but the transitions become independent of molecular weight when the polymer chains contain more than 30–50 repeat units (see Table 1, **I-5** for different monomer: initiator ratios). This effect has also been confirmed by other authors, and it may be considered to be a general feature for ROMP-derived SCLC polynorbornenes [39, 40]. Increasing the length of spacer chains also induces side chain crystallization in low molecular oligomers. This effect takes place in oligomers **I-n** with *n*>9 and with a degree of polymerization less than 20, and in all samples of **I-11** and **I-12**. Side-chain crystallization is strongly kinetically controlled and is therefore suppressed by increasing molecular weight; so the enthalpy of melting and therefore the extent of crystallinity decreases as the polymer chain length increases. The effect of side chain crystallization was not observed for polymers with shorter spacer lengths, except for the two shortest oligomers of **I-5**. The authors could not explain why only these oligomers show side chain crystallization. In two blends of low molecular oligomers with higher molecular polymers (**I-6, I-7**), the authors showed that the thermal transitions of polydisperse samples have also sharp transitions. Polydispersity has no influence on the peak widths of the transitions if the component polymers are isomorphic over the temperature range at which the transition occurs [38].

Changing the methoxy-substituent of the biphenyl entity to a cyano-group produces a significant influence on the mesophases. Komiya et al. report the synthesis of norbornene derivatives with the (4'-cyanobiphenyl-4-yl)oxy-mesogen linked to the norbornene entity with different spacer lengths via an ester (**II-n**, *n*=3–7, 9–12) as anchoring group, see Fig. 4 [39]. Polymerization was carried out with the Schrock-type initiator **1**. Polymers of **II-3** were amorphous, whereas all other polymers (*n*=4–7, 9–12) displayed enantiotropic nematic mesophases (Table 1). Analogous to poly-(**I-n**), transition temperatures increased with molecular weight, but became independent of molecular weight at 30–50 repeat units. In contrast to the methoxy analogues, no odd-even effect was found for the isotropization temperature, which was almost independent of spacer length, whereas the glass transitions decreased with increasing spacer lengths. No tendency towards side chain crystallization, even for oligomers

Table 1 Polymerization of monomers I–n, II–n and III–n with initiator 1 (data given for monomer to initiator ratio=1:100; for other ratios see references)

Monomer	[1]:[1–n]; [mol]:[mol]	PDI	M_n	DP	Phase transitions (°C) and enthalpy changes (kJ mru^{-1}, in parentheses)[a]	Reference
I–2	1:100	1.05	37,306	102	g 74 n 93 (0.65) i	[38]
I–3	1:100	1.20	43,993	116	g 61 n 88 (0.64) i	
I–4	1:100	1.05	52,857	135	g 51 n 96 (1.13) i	
I–5	1:100	1.12	61,247	151	g 44 n 88 (1.34) i	
I–5	1:50	1.06	26,386	65	g 43 n 87 (1.39) i	
I–5	1:20	1.11	8185	20	g 40 n 80 (1.16) i	
I–5	1:10	1.12	4963	12	g 51 k 74 (–3.64) k 90 (5.93) i	
I–5	1:5	1.12	3490	9	g 46 k 66 (–1.92) k 87 (6.85) i	
I–6	1:100	1.17	61,022	145	g 43 n 95 (1.43) i	
I–7	1:100	1.17	34,089	78	g 38 n 95 (2.12) i	
I–8	1:100	1.23	46,386	103	g 30 n 91 (1.86) i	
I–9	1:100	1.07	47,949	104	g 29 n 91 (2.24) i	[37]
I–10	1:100	1.26	62,683	132	g 23 n 87 (1.85) i	
I–11	1:100	1.23	35,086	72	g 40 k 49 (1.01) s 84 (3.56) i	
I–12	1:100	1.28	49,568	98	g 23 k 61 (2.99) s 89 (2.70) i	
II–3	1:100	1.23	65,700	176	g 78 i	[39]
II–4	1:100	1.17	52,000	134	g 66 n 103 (1.58) i	
II–5	1:100	1.18	42,800	107	g 54 n 92 (1.58) i	
II–6	1:100	1.27	64,600	155	g 48 n 92 (1.84) i	
II–7	1:100	1.27	124,800	290	g 45 n 95 (2.15) i	
II–9	1:100	1.20	91,500	200	g 41 n 96 (2.80) i	
II–10	1:100	1.27	44,800	95	g 36 n 94 (3.23) i	
II–11	1:100	1.20	99,500	205	g 43 n 97 (3.12) i	
II–12	1:100	1.24	52,900	106	g 27 n 85 (3.46) i	
III–4	1:100	1.15	44,700	118	g 54 k 108 (8.45) i	
III–5	1:100	1.14	79,500	202	g–k 92 (7.60) i	
III–6	1:100	1.12	49,400	122	g–k 94 (9.15) i	
III–5[b]	1:50	2.8	31,000	–	smectic	[21]

[a] Phase transitions are given for the second heating run (DSC); mru: molecular repeating unit;
[b] initiator: $WCl_4[OC_6H_3(C_6H_5)2]_2$-$PbEt_4$ (1:2).

with relatively long spacers, could be detected. These effects were attributed to dipole-dipole repulsions between adjacent (4'-cyano-biphenyl-4-yl)-oxy-mesogens, disturbing an ordered arrangement of the side chain. The normal antiparallel arrangement of cyanobiphenyloxy-mesogens cannot be adopted because of the linkage to the polymer chain.

In the same work, norbornene derivatives with a (4'-methoxy-biphenyl-4-yl)-oxy mesogen linked by an ether group III–n (n=4–6, see Fig. 4) were synthesized. Surprisingly, in contrast to ester-linked polymers (I–n) [38], all of the poly-(III–n) polymers gave highly crystalline polymers (Table 1) caused by side chain crystallization, even in polymers with relatively short spacers (n=4–6).

This effect was attributed to the higher flexibility of the ether linkage, as there is no rotational restriction of the C–O–C bond in contrast to the ester group, whose asymmetry is believed to disturb the side chain crystallization.

In contrast to this work, polymerization of **III-5** with the tungsten initiator $WCl_4[OC_6H_3(C_6H_5)_2]_2$-$PbEt_4$ (1:2) led to polymers with a smectic mesophase (see Table 1, last entry) [21].

Gangadhara et al. have linked the cyanobiphenyl mesogen via a dicarboximide-group to an oxanorbornene ring system **VI-n**, n=2–8 (see Fig. 4). Polymerization was carried out with Schrock type initiator **4**. The dicarboximide linkage probably hindered the formation of LC phases; even the introduction of relatively long spacers between the polymer backbone and the mesogen did not lead to liquid crystalline monomers or polymers [41].

2.1.2
Influence of Mesogen Density

A double mesogen density is easily achieved by using disubstituted norbornene monomers of type **IV-n** and **V-n** (Fig. 4).

Our group investigated the influence of the spacer length of 2,3-disubstituted norbornene derivatives with cyanobiphenyl-mesogenic units [42] and with methoxybiphenyl-mesogenic units [43] on the mesophases. *Exo,endo*-bis-[(4'-cyanobiphenyl-4-yl)oxy-*n*-alkyl]norborn-5-ene-2,3-dicarboxylates with different alkyl spacer lengths (**IV-n**, n=2–12) were synthesized in a straightforward manner.

Polymers were prepared by ROMP using initiator **1** in an initiator to monomer ratio of 1:100. Because of the high reactivity of the initiator towards solvent impurities and moisture, the DP of almost all polymers exceeds the theoretical value (see Table 2). All polymers showed thermotropic liquid crystalline phases. A strong decrease in the glass transition temperature T_g with increasing spacer length was observed. It changed from about 90 °C for poly-(**IV-2**) to 20 °C for poly-(**IV-12**) in an exponential manner, whereas the isotropization temperatures remained within a narrow temperature range, from 107 to 118 °C. Polymers with spacers of 2–7 methylene units showed enantiotropic nematic phases, and in contrast to the monosubstituted cyanobiphenyl polynorbornene derivatives (**II-n**) [39], their isotropization temperatures exhibited a strong odd-even effect. The higher number of mesogens shifted the clearing temperatures of the LC-phases to higher temperatures (compared to poly-(**II-n**), with values around 90–100 °C, see Table 1). Polymers with longer spacers (n=8 to 12) showed a smectic A phase without a significant odd-even effect. With a spacer length of six methylene groups, the mesophase type for poly-(**IV-6**) was strongly dependent on the stereochemistry in the main chain – the Z/E ratio of the double bonds – as will be discussed in the next section. Poly-(**IV-6**), containing almost all Z-isomers, formed a smectic A phase.

Another set of polymers were synthesized from (±)-*exo,endo*-bis[ω-(4'-methoxybiphenyl-4-yloxy)alkyl]bicyclo[2.2.1]-hept-5-ene-2,3-dicarboxylates,

Table 2 Physico-chemical data for poly-(IV–n) and poly-(V–n)

Monomer	Initiator	PDI	M_n	DP	Phase transitions (°C) and enthalpy changes (kJ mru^{-1}, in parentheses)[a]	Reference
IV–2	1	1.68	99,900	159	g 95.4 n 115.1 (1.9) i	[42]
IV–3		1.22	81,400	124	g 80.8 n 107.3 (1.7) i	
IV–4		1.34	59,400	87	g 62.4 n 114.0 (1.8) i	
IV–5		1.66	108,100	153	g 56.7 n 109.5 (2.1) i	
IV–6		1.44	103,300	140	g 52.7 n 118.4 (2.2) i	
IV–7		1.35	139,300	182	g 46.5 n 112.9 (2.1) i	
IV–8		1.58	294,700	372	g 39.0 s_A 114.8 (5.4) i	
IV–9		1.34	209,000	255	g 30.9 s_A 116.3 (4.8) i	
IV–10		1.31	111,500	131	g 27.2 s_A 114.9 (5.9) i	
IV–11		1.39	189,000	215	g 25.7 s_A 118.0 (7.2) i	
IV–12		1.38	149,500	165	g 24.9 s_A 116.1 (7.2) i	
V–4	2	1.20	27,000	–	k 144.9 (19.4) s 153.7 (2.9) i	[43]
V–5		1.36	55,000	–	k 127.0 (16.1) s 140.7 (4.0) i	
V–6		1.46	53,000	71	k 122.6 (14.7) s 150.5 (4.4) i	
V–6		1.26	27,500	29	k 123.2 (14.4) s 147.7 (4.5) i	
V–6		1.29	22,900	24	k 123.3 (15.4) s 145.3 (4.6) i	
V–6		1.35	18,800	19	k 123.3 (16.5) s 137.7 (4.9) i	
V–6		1.23	9400	10	k 124.0 s 132.2 (–) i	
V–7		1.19	23,000	–	k 103.0 (12.1) s 140.0 (4.1) i	
V–8		2.13	99,000	–	k 130.7 (29.2) s 152.9 (6.0) i	
V–9		2.75	144,000	–	k 124.29 (25.8) s 148.1 (6.8) i	
V–10		2.52	80,000	–	k 134.3 (32.8) s 150.4 (8.3) i	

[a] Phase transitions are given for the second heating run (DSC).

with spacer lengths varying from four to ten methylene units (V–n, n=4–10), using initiator 2 (see Table 2) [43]. All of these polymers showed smectic mesophases. Independently of the spacer length, no glass transition was detected; all of the polymers had a crystalline-smectic transition. Some polymers showed additional thermal transitions with low enthalpy changes, but no identification of the type of transition was possible. A dependency of the liquid crystalline transition temperature on the molecular weight was observed for poly-(V–n) (Table 2). Transition temperatures were roughly constant for polymers consisting of more than 30 repeating units. In contrast, the enthalpy changes of these transitions were nearly independent of the molecular weights once the polymers contained more than ten repeating units, but they were strongly dependent on spacer length, with a minimum at n=6.

In both cases, the increase in the number of mesogenic units led to a preferential formation of smectic phases, especially with methoxy-biphenyl as mesogen.

Danch et al. studied the effect of thermal history on the assignment of the glass transition event associated with the biaxial orientation of the smectic phase of **IV-n** with $n=6$ or 11, using DSC [44]. In further works, relaxation phenomena of these polymers were studied by dynamic mechanical thermoanalysis and dielectric thermoanalysis measurements [45]. The rheological behaviors of solutions of low and high molecular poly-(**IV-n**) ($n=5, 9$) were investigated by Wewerka [46].

2.1.3
Influence of Tacticity

Ungerank et al. investigated the influence of different molybdenum Schrock-type initiators **1, 2** and **3** (see Fig. 2) on the polymerization of (±)- and (−)-*exo,endo*-bis[4′-cyanobiphenyl-4-yl)oxyalkyl]norborn-5-ene-2,3-dicarboxylates [(±)-**IV-n** and (−)-**IV-n**] [47]. The initiator had a strong influence on the *Z/E* ratio of the double bond in the polymer chain and therefore also on the tacticity of the polymer chain. Using racemic 2,3-disubstituted norbornene derivatives, such as (±)-**IV-n,** the two adjacent monomers (diads) can in principle form eight different stereoisomers, as depicted in Fig. 5 [48].

None of these three initiators formed one specific diad; only a certain percentage of *cis* double bonds (σ_c) or of tacticity (*iso-* or *syndio-tactic*) was

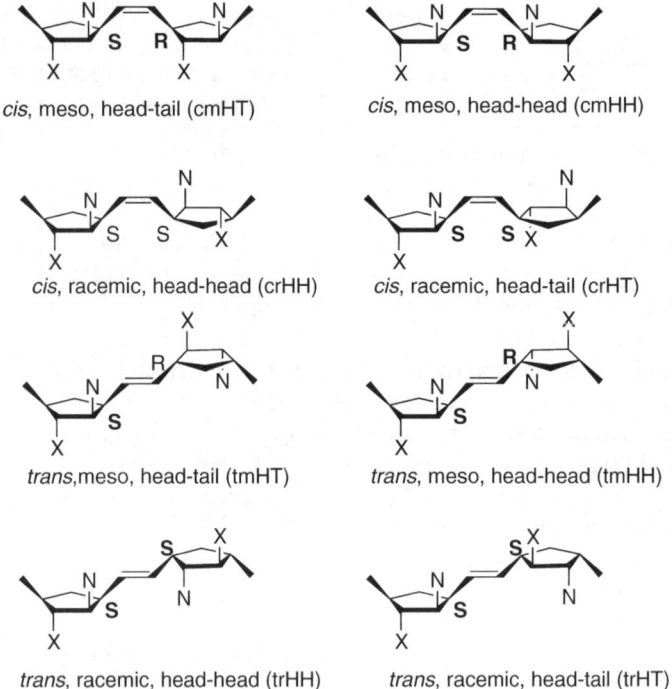

Fig. 5 Eight possible diads for a (±)-*trans*-2,3 disubstituted polynorbornene chain; the X substituent is derived from the *exo*-position, N from the *endo* position in the monomer

Table 3 Percentage of Z-double bonds in a poly(norbornene) chain

Polymer	Polymerized with		
	1	3	2
(±)-poly-**NbdMe**	20	54	78
(−)-poly-**NbdMe**	26	56	80

achieved. The tacticity of the polynorbornene main chain was first determined by various NMR techniques for model polymers (poly-{bicyclo[2.2.1]hept-5-ene-2,3-dicarboxylic acid dimethyl ester}, **NbdMe**). In the case of poly-(**IV–n**), the resolutions of the carbon spectra were too low for detailed microstructure investigations. Using the pure enantiomers (−)-**NbdMe**, the number of potential diads could be reduced by half. The percentage of Z-double bonds strongly increased from 20%, in polymers obtained with initiator **1**, to around 55% with initiator **3** and 80% with initiator **2** (Table 3).

Data for the polymers poly-(**IV–6**) and poly-(**IV–11**) are presented in Table 4. In general, the increase in the number of Z-double bonds led to higher isotropization temperatures but lower T_g temperatures. All polymers with shorter spacer chains (n≤ 5) showed a nematic mesophase, whereas all polymers with spacer lengths n≥7 gave a smectic A phase, independent of the initiator used. Polymers of **IV–6** were an interesting case, since they were prepared with initiator **1** and so had a low percentage of *cis* bonds. They formed nematic mesophases, but polymers synthesized with initiators **2** and **3** showed smectic A phases.

This means that the formation of the mesophase was sensitive to the initiator used. In the case of a *trans*-diad, the length of one repeating unit is 6.2 Å,

Table 4 SEC and DSC data for selected polymers from [47]

Polymer	Initiator	PDI	M_n	DP	Phase transitions (°C) and enthalpy changes (kJ mru^{-1}, in parentheses)[a]
(±)-**pIV–6**	1	1.44	103,300	140	g 52.7 n 118.4 (2.2) i
(±)-**pIV–6**	3	1.52	61,300	87	g 51.2 s$_A$ 123.9 (3.7) i
(±)-**pIV–6**	2	1.10	61,300	83	g 45.5 s$_A$ 139.5 (3.9) i
(±)-**pIV–11**	1	1.39	189,000	215	g 25.7 s$_A$ 118.0 (7.2) i
(±)-**pIV–11**	3	1.20	77,700	88	g 24.7 s$_A$ 125.9 (8.7) i
(±)-**pIV–11**	2	1.12	91,400	104	g 23.4 s$_A$ 141.5 (9.0) i
(−)-**pIV–11**	1	1.88	124,400	141	g 19.5 s$_A$ 118.0 (6.9) i
(−)-**pIV–11**	3	1.54	117,900	134	g 21.0 s$_A$ 131.6 (7.5) i
(−)-**pIV–11**	2	1.48	154,700	175	g 18.5 s$_A$ 148.6 (8.8) i

[a] Phase transitions are given for the second heating run (DSC).

while it is only 5.2 Å for the *cis*-diad. If the space requirement of a mesogenic unit in a smectic phase – about 4.5 Å – is taken into account, a *cis*-diad meets this requirement much better than a *trans*-diad does. A *trans*-diad is too long, which causes kinks in the backbone and in the smectic A layer. Therefore, the formation of smectic phases is disfavored for polymers synthesized with initiator 1 compared to 2.

Differences in tacticity were also reflected by thermal data. While the isotropization temperature of (–)-poly-(**IV–11**) and (±)-poly-(**IV–11**), synthesized using initiator 1, stayed approximately unchanged, the isotropization temperatures for the chiral liquid crystalline polymers shifted to higher values when initiators 2 or 3 were used. The difference was up to 7 °C. If the decreased glass transition temperatures (T_g) for the chiral analogues were taken into account, the temperature range of the liquid crystalline phase was broadened by up to 12 °C. This means that a certain diad must be responsible for this behavior. The authors assumed that the diad cmHT was most suitable one for the formation of stable liquid crystalline phases in poly(norbornene) main chains.

The influence of diluting on mesogenic units with a non-LC-monomer was investigated by synthesizing copolymers of **IV–6** and **NbdMe** in different ratios with initiator 2. The isotropization temperature decreased with increasing amounts of **NbdME**. Above a critical value (around 50% for this system), no LC-phase was observed [47].

2.1.4
Magnetic Orientation of SCLCPs

One of the most interesting features of SCLCPs is related to the fact that a liquid crystalline phase can be orientated and frozen by cooling it to below the glass transition temperature. It is therefore necessary to drive SCLCP systems from microscopic self-organized mesophases to macroscopic order.

Boamfa et al. investigated macroscopic orientation of disubstituted cyanobiphenyl poly(norbornene) **IV–n** ($n=3, 5$) using different magnetic fields of up to 20 T. Both the degree of orientation and optical properties were compared with a cyanobiphenyl-substituted acrylate polymer with a spacer of four methylene units [49, 50].

Birefringence and optical transmission data were collected simultaneously while cooling the SCLCPs in the magnetic field down from the isotropic phase through nematic phase to below T_g. The samples obtained were highly transparent and show uniaxial orientation of the mesogen moieties along the magnetic field direction.

The degree of polymerization was related to the minimum field needed to induce a significant effect. For DP=20 the minimum field was 6 T, whereas for DP=30 it was at least 10 T, rising to 19 T for DP=40. For a shorter spacer length ($n=3$), no induced effect was observed up to 20 T, which can be explained by the reduced mobility of the mesogenic moieties, resulting in reduced rotation towards the field director.

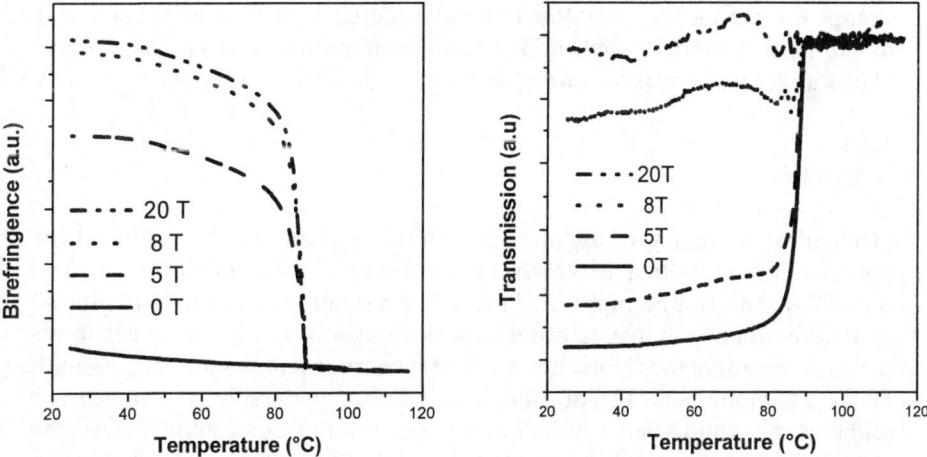

Fig. 6 Magnetic field-induced birefringence and transmission for a poly-(**IV–5**-COEN) sample in various magnetic fields

Recently, the orientation behavior of poly-(**IV–5**) was compared to both poly-(**IV–4**) and a statistical copolymer poly-**IV–5**-COEN (COEN=cyclooctene) [51]. The experiments showed that, compared to poly-(**IV–5**), samples of poly-(**IV–5**-COEN) were easier to orientate, a finding that was attributed to the more flexible chain. The dependence of the magnetic field and the temperature on the degree of orientation, measured by the birefringence and the optical transparency, is shown in Fig. 6.

After orientation at high magnetic field, the previously turbid samples were optically transparent, as can be seen in Fig. 7. Poly-(**IV–4**) lost some degree of orientation after removing the sample from the magnetic field, leading to a slight turbidity (see the glass slide denoted C4 in Fig. 7). The other two samples were found to be stable for several weeks.

Finally, this method possesses a unique set of advantages compared to other orientation methods (electric field, flow field, mechanical field): it is clean and contact-free, it does not create any electrodynamic instabilities, and it does not

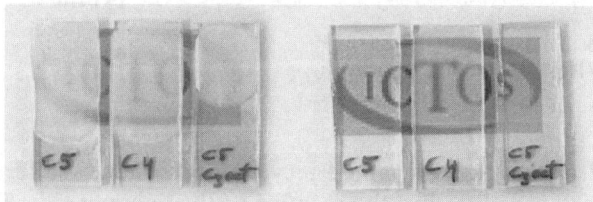

Fig. 7 Samples of poly-(**IV–5**) ("C5"), poly-(**IV–4**) ("C4"), and copolymer poly-(**IV–5**-COEN) ("C5-Cyoct") before (left) and after (right) orientation in a magnetic field

change the sample thickness. Such highly ordered SCLCPs may be applicable to the field of optical applications (nonlinear optics, retardation-polarizer plates, alignment layers for low molecular weight liquid crystals).

2.1.5
LC Elastomers

Mechanical alignment of the mesogens in the LC-state is only possible if the polymers are crosslinked to form an elastomer. For the poly(norbornene)s poly-(**V–n**) this was accomplished by adding a small amount (up to 10 mol %) of a difunctional polymer. In order to achieve a good crosslinking yield, it was necessary to adopt the chain between the two norbornenes to the spacer length of the LC monomers. In addition, a stiff link – preferably a phenylene or biphenylene group – was required in the center of the molecule, otherwise the effectiveness of the crosslinking was reduced dramatically (most probably due to the formation of loops). In a polymer with a low T_g-gained by copolymerizing the norbornene compound with cyclooctadiene, forming poly(butadiene) sequences, for example–mesogen orientation could be accomplished by mechanical stretching of the "rubber band". In the stretched mode the polymer was very transparent, while it was opaque in the unstrained state [52].

2.1.6
SCLCPs with Different Backbones

ROMP of monomer **VII** (see Scheme 1) was successfully carried out by our group with initiator **2** [53]. Due to the different ring structure – cyclooctene was

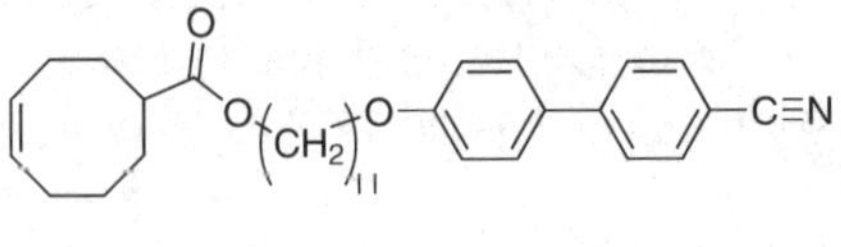

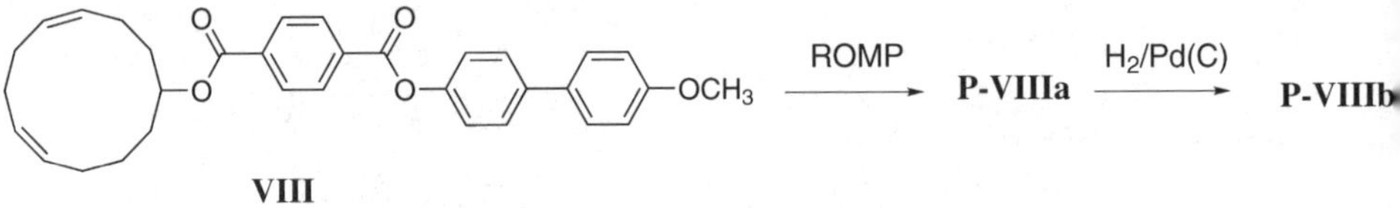

Scheme 1

used instead of the highly strained bicyclic norbornene – a large excess of monomer was needed to shift the reaction to the polymer side. The polymer had a T_g<0 °C, and showed a nematic LC phase, and was therefore considered a good candidate for LC elastomers (Table 5).

Table 5 Molecular weight characteristics and thermal transitions of poly-VII and poly-VIII

Polymer	Initiator	PDI	M_n	Phase transitions (°C) and enthalpy changes (kJ mru^{-1}, in parentheses)[a]	Reference
p-VII	2	6.8	38,000	c –5.3 n 53.2 (1,4) i	[53]
p-VIIIa	WCl$_4$(OAr)$_2$	2	10,500	g 62 s 116 i	[54]
p-VIIIb	–	–	–	g 61 s 113 i	

[a] Phase transitions are given for the second heating run (DSC).

Cho et al. described the synthesis and polymerization of 4,8-cyclododeca-dien-1-yl-(4'-methoxy-4-biphenyl) terephthalate VIII [54, 55]. Polymerization was carried out with WCl$_4$(OAr)$_2$/PbEt$_4$. The double bonds in the polymer backbone were subsequently hydrogenated with H$_2$/Pd(C), leading to a SCLCP with a fully saturated hydrocarbon backbone. This polymer system had a very flexible polymer backbone but a stiff connection between the main chain and the mesogenic unit. The distance between two adjacent side chains was about 12 methylene units. This very flexible main chain allowed the polymer to organize into a LC mesophase. Both polymers – the unsaturated and the saturated – showed smectic liquid crystalline mesophases with almost the same transition temperatures (see Table 5).

2.1.7
Comparison of Cyanobiphenyl-Based SCLCPs Prepared by Different Methods

In this chapter we want to discuss the correlation of the mesophase behavior of a cyanobiphenyl-based SCLCP with its backbone structure. As shown before, the backbone structure, the spacer lengths, and the mesogen density per repeat unit have great influence on the LC mesophase evolved. Figure 8 shows some examples of backbone structures bearing the cyanobiphenyl-moiety that have been reported in literature. The above-mentioned ROMP-derived polymers poly-(II–n) [39], poly-(IV–n) [42, 47], poly-(VI–n) [41], and poly-(VII–n) [53] will be compared with each other and with acrylate-based [56–59], siloxane-based [60] and vinylcyclopropane-based systems [61]. The detected mesophases and their transition temperatures are summarized in Table 6.

For polymers in which the mesogen is separated by a spacer of six methylenes units from the polymer backbone, it is obvious that the more rigid poly(norbornene)s favor nematic liquid crystalline phases. Poly-(VI–6) with the rigid and bulky 2,5-dimethine oxacyclopentane-3,4-dicarboximide unit in the main chain does not show liquid crystalline behavior (Table 6, entry 12). The more flexible poly-(II–6) backbone allowed the formation of a nematic mesophase. If the mesogen density was increased, as realized in poly-(IV–6), the isotropization temperature was found to be 26 °C higher than that for poly-

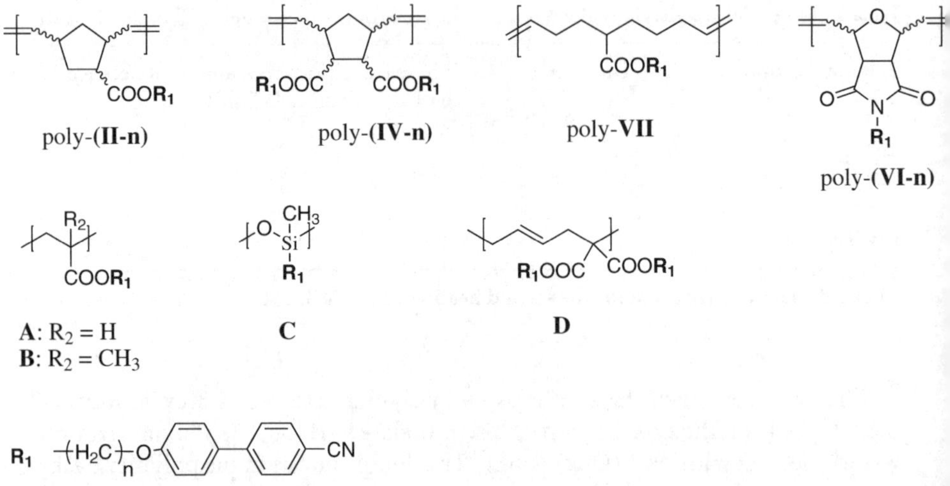

Fig. 8 Polymer backbone structures of SCLCPs based on cyanobiphenyl mesogens

Table 6 Comparison of SCLCPs with different main chains based on the cyanobiphenyl-mesogen

Entry	Polymer	Initiator	n	Phase transition	Reference
1	poly-(**II-6**)	1	6	g 48 n 92 i	[39]
2	poly-(**II-11**)	1	11	g 43 n 97 i	[39]
3	poly-(**IV-6**)	1	6	g 52.7 n 118.4 i	[42]
4	poly-(**IV-6**)	3	6	g 51.2 s_A 123.9 i	[47]
5	poly-(**IV-6**)	2	6	g 45.5 s_A 139.5 i	[47]
6	(±)-poly-(**IV-11**)	1	11	g 27.2 s_A 114.9 i	[42]
7	(±)-poly-(**IV-11**)	3	11	g 24.7 s_A 125.9 i	[47]
8	(±)-poly-(**IV-11**)	2	11	g 23.4 s_A 141.5 i	[47]
9	(−)-poly-(**IV-11**)	1	11	g 19.5 s_A 118.0 i	[47]
10	(−)-poly-(**IV-11**)	3	11	g 21.0 s_A 131.6 i	[47]
11	(−)-poly-(**IV-11**)	2	11	g 18.5 s_A 148.6 i	[47]
12	poly-(**VI-6**)	4	6	g 77.4	[41]
12	poly-**VII**	2	11	k −5.3 n 53.2 i	[53]
13	A		6	g 32 n_{Re} 80 s_A 124 n 132 i	[56]
14	A		11	g 25 s_C 30 s_A 145 i	[57]
15	B		6	g 55 s 100 i	[58]
16	B		11	g 40 s_A 121 i	[59]
17	C		6	g 14 s_A 166 i	[60]
18	C		11	g −1 s_C 48 s_A 157 i	[60]
19	D		6	g 39 s 151 i	[61]

(II–6) (Table 6, entries 1 and 3). This indicates an enhanced stability, and consequently a higher state of order for the liquid crystalline phase. This is confirmed by the fact that polymers with a higher number of Z-double bonds (Table 6, entries 4 and 5) show a smectic mesophase. Smectic mesophases were also found in polymers with poly(acrylate) **A**, poly(methacrylate) **B**, poly(siloxane) **C**, and poly(vinylcyclopropene) **D** backbones (Table 6 entries 13, 15, 17, 19). The isotropization temperatures of these polymers were approximately the same.

For a spacer length of 11 methylene units, all polymers showed smectic mesophases except for poly-(II–11) and poly-(VII). Even the flexible poly-(cyclooctene) main chain prevented a smectic mesophase. Compared to all of the other polymer architectures, poly-(II–11) and poly-VII present the lowest ratios of mesogens to atoms in the main chain. It can therefore be assumed that smectic phases will only be formed when there is sufficient mesogen density. For the norbornene chain, it is notable that a high Z/E-ratio and a high tacticity increased the stability of the smectic A phase.

2.1.8
SCLCP Block Copolymers

Statistical copolymerization of SCLC-monomers with non-liquid crystalline monomers leads to dilution of the mesogenic units in the polymer, and (below a critical value) to the loss of the LC behavior of the polymer [47].

Block copolymers with well-defined segments often show microphase-separated morphologies (such as lamellar layers, hexagonal ordered cylinders, and micelle formation). If we use SCLCP blocks together with non-liquid crystalline segments, the mesophases are formed within one of the separated microdomains. If the non-SCLCP block has a higher T_g than the phase transition temperature of the mesophase, the amorphous block should physically support the SCLCP microdomains, forming a "self-supported" SCLCP system.

Komiya et al. described the living ROMP synthesis of AB-type block copolymers that contain side chain liquid crystalline polymer blocks and amorphous blocks [62]. Norbornene (**NBE**), 5-cyano-2-norbornene (**NBCN**) and methyltetracyclododecene (**MTD**) were used for the amorphous polymer blocks, while **I-n** (n=3, 6) were used for the SCLCP block (see Fig. 9). Initiator **1** was used for the ROMP. Block copolymers with monomer ratios from 75/25 to 20/80 (amor-

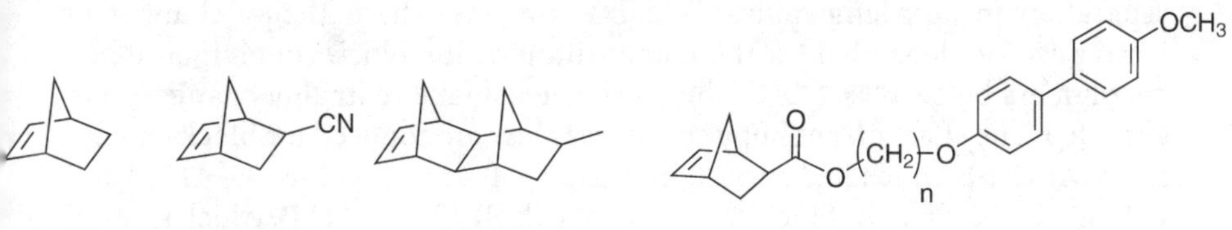

NBE   **NBECN**   **MTD**   **I-n, n = 3, 6**

Fig. 9 Monomers used for block copolymerization in [62]

Table 7 Properties of block copolymers containing **I-n** and norbornene(NBE), methylcyclododecene (MTD) or cyanonorbornene (NBECN)

Monomer 1	Monomer 2 (found)	Block ratio	PDI	M_n	Phase transitions (°C)
NBE	I-6	75/25	1.11	54,100	g 42 n 97 i
NBE	I-6	49/51	1.17	27,100	g 38 n 94 i
NBE	I-6	30/70	1.25	38,000	g 38 n 91 i
NBE	I-6	20/80	1.11	32,700	g 38 n 93 i
NBE	I-3	79/21	1.06	88,900	g 44 g 58 n 84 i
NBE	I-3	53/47	1.10	37,600	g 39 g 55 n 92 i
NBE	I-3	33/67	1.13	64,200	g 39 g 59 n 86 i
NBE	I-3	20/80	1.16	50,800	g 36 g 60 n 87 i
MTD	I-6	67/33	1.08	28,500	g 41 n 85 g 208 i
MTD	I-6	47/53	1.18	42,900	g 41 n 92 g 207 i
NBECN	I-6	70/30	1.17	113,000	g 45 n 87 g 112 i
NBECN	I-6	52/48	1.13	145,900	g 43 n 93 g 114 i
NBECN	I-6	38/62	1.16	79,300	g 43 n 91 g-[a] i
NBECN	I-6	21/79	1.16	47,000	g 45 n 91 g-[a] i
NBECN	I-3	72/28	1.09	66,200	(g–n–)[a] g 112 i
NBECN	I-3	54/46	1.17	108,700	g 67 n 86–91 g 119 i
NBECN	I-3	33/67	1.08	61,800	g 67 n 82 g-[a] i
NBECN	I-3	23/77	1.15	46,600	g 67 n 82 g–i
–	I-6	0/100	1.15	24,100	g 42 n 93 i
–	I-3	0/100	1.12	20,200	g 61 n 86 i
NBE	–	100/0	1.07	47,674	g 43 i
MTD	–	100/0	1.07	19,995	g 214 i
NBECN	–	100/0	1.06	82,079	g 116 i

[a] Could not be detected.

phous/SCLCP) and narrow molecular weight distributions (PDI=1.06–1.25) were obtained in high yields (Table 7; the values of the enthalpy changes are given in the reference). The glass transition temperatures of each block and the isotropization temperatures of the mesophases were found to be independent of the composition of the block copolymer and identical to those of the respective homopolymers. Except for the block copolymer with the highest content of **NBECN**, which was amorphous, the LC mesophases and the transition temperatures were independent of the block size. These findings suggest phase separation in both amorphous and LC domains. The enthalpy changes of isotropization depended on the composition of the block copolymer. Block copolymers below 60% SCLCP block showed smaller enthalpy changes than expected. The effect on enthalpy change was less pronounced for block copolymers with **NBE**, which had almost the same T_g (40–43 °C) as **I-6** (35–42 °C) and **I-3** (57–61 °C), than for block copolymers with **NBCN** and **MTD**, which showed glass transition temperatures above the isotropization temperature of **I-n** (poly-**MTD**: T_g=214 °C, poly-**NBECN**: T_g=116 °C). The amorphous glassy domains of poly-**NBECN** and poly-**MTD** hindered the orientation of the LC phases and moreover had a higher impact on the mesophase of **I–3** with a shorter spacer length. On the other hand, the **NBE**-block was already above the glass transition and therefore more flexible.

Block copolymers consisting of a smectic SCLCP-block and a partially crystalline apolar block were synthesized via ROMP of **IV–n** with cyclooctene and initiator **1** or **2** [63]. The block copolymers also formed smectic liquid crystalline mesophases and showed lamellar phase-separation.

2.2
Terminally Attached Mesogens of Various Kinds

The influence of the rigidity of the backbone on the mesophases of *p*-nitrostilbene-containing polymers was the subject of a study by Maughon et al. [64]. The rigid poly(norbornene) backbone was compared with the flexible poly(butadiene) chain. For this purpose, *p*-nitrostilbene was linked by an ester group to the norbornene bicycle, **IX–n** ($n=6, 8, 10, 12$), and to the cyclobutene, **X–n**, $n=6, 8, 10, 12$, respectively (Fig. 10). Polymers were synthesized with initiator **6**; SEC and DSC data of the resulting polymers are summarized in Table 8.

Stabilization of the mesophase was observed as the degree of polymerization was increased. The T_g values of the poly(norbornene)-polymers were about 30 °C higher than those of the poly(butadiene) polymers. Both polymers showed similar isotropization temperatures, but they differed substantially in their liquid crystalline behaviors. Poly-(**IX–n**)s with a poly(norbornene) backbone exhibited textures typical of nematic mesophases, whereas the poly(butadiene)-based polymers poly-(**X–n**) displayed textures representative of smectic A mesophases. The more flexible backbone of poly(butadiene) allowed a higher order of alignment of the mesogenic units, resulting in the more ordered liquid crystalline smectic A phase.

One example of a 1:1 AB block copolymer with poly-(**IX–8**) and poly-(**X–8**) blocks showed a smectic A phase; the mesophase of the homopolymer of **X–8**. The more ordered smectic A phase dominated the nematic phase of the polynorbornene block.

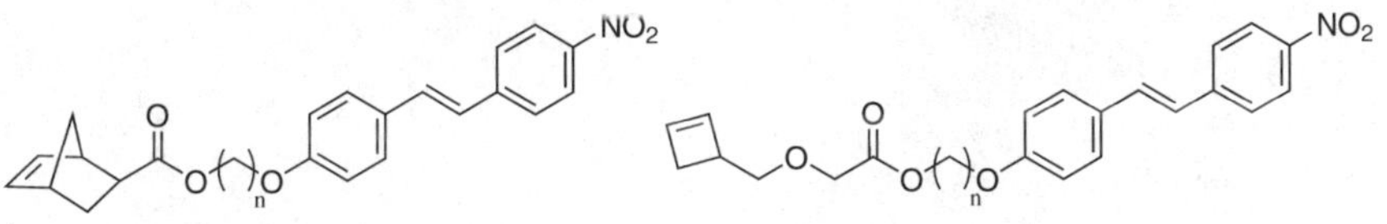

IX-n
n=6, 8, 10, 12

X-n
n=6, 8, 10, 12

Fig. 10 Chemical structures of monomers based on *p*-nitrostilbene

Table 8 SEC and DSC data for poly-(**IX–n**) and poly-(**X–n**)

Monomer	Monomer:initiator	PDI	M_n	Phase transitions (°C)[a]
IX–6	25	1.08	13,400	g 64 n 121 i
IX–8	25	1.08	13,000	g 52 n 117 i
IX–10	25	1.11	22,000	g 50 n 108 i
IX–12	25	1.10	20,800	g 44 n 108 i
IX–8	5	1.11	7300	g 46 n 108 i
IX–8	10	1.09	9800	g 50 n 114 i
IX–8	50	1.07	18,000	g 51 n 118 i
IX–8	100	1.08	23,200	g 49 n 121 i
X–6	25	1.16	33,300	g 31 s 104 i
X–8	25	1.15	35,400	g 25 s 104 i
X–10	52	1.16	31,900	g 23 s 111 i
X–12	25	1.14	38,500	g 21 s 108 i
X–8	5	1.13	15,000	g 23 s 74 i
X–8	10	1.11	21,600	g 21 s 86 i
X–8	50	1.27	57,000	g 15 s 105 i
X–8	100	1.38	89,100	g 14 s 107 i

[a] Phase transitions are given for the second heating run (DSC).

Very recently, the azobenzene norbornene derivative **XI** (Scheme 2) was successfully prepared and polymerized with initiator **5** and **9**, and it showed a nematic mesophase. The polymer was photo-switchable due to photo *cis/trans*-isomerization of the N=N double bond [65]. This is the first step towards ROMP-derived materials that may be used for optical data storage.

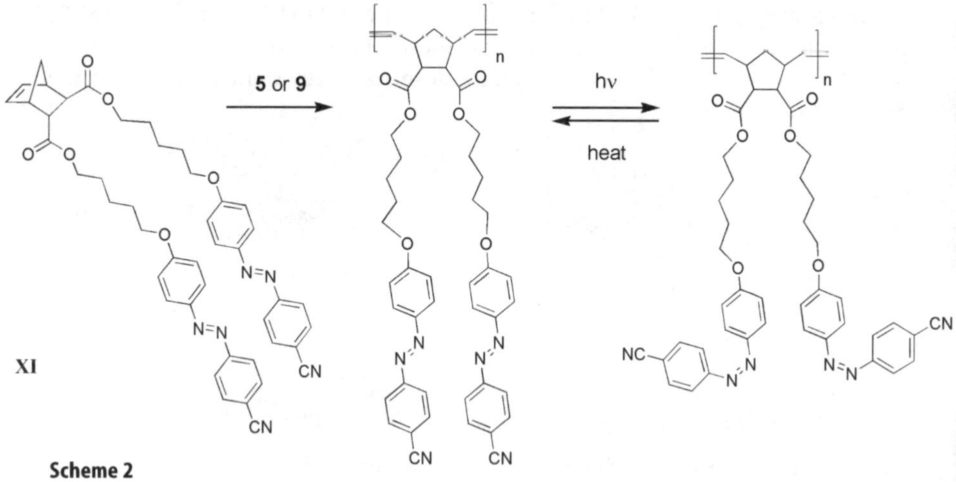

Scheme 2

2.2.1
Block Copolymers

The introduction of perfluorinated groups generally favors microphase separation due to the immiscibility of fluorocarbons with hydrocarbons [66]. Norbornene derivatives with perfluorinated endgroups in the side chain were prepared by Wewerka et al. [67]. Monomer **XII** contained a relatively long $(CF_2)_8$-chain, separated via a long spacer (11 methylene-groups) from the norbornene, whereas monomer **XIII** has two relatively short $(CH_2)_2(CF_2)_4$-side chains (Fig. 11). Homopolymers and block copolymers were synthesized with one fluorinated monomer (**XII** or **XIII**) and one non-fluorinated non-liquid crystalline monomer (**NBDE** or **COEN**) with the Schrock-type initiators 4 and 5, respectively, leading to microphase-separated block copolymers. Table 9 and Table 10 summarize the physico-chemical properties of the homopolymers and block copolymers.

Among the homopolymers, poly-**XII** showed smectic mesophases, poly-**XIII** and poly-**NBDE** were amorphous, and poly-**COEN** was semi-crystalline. The

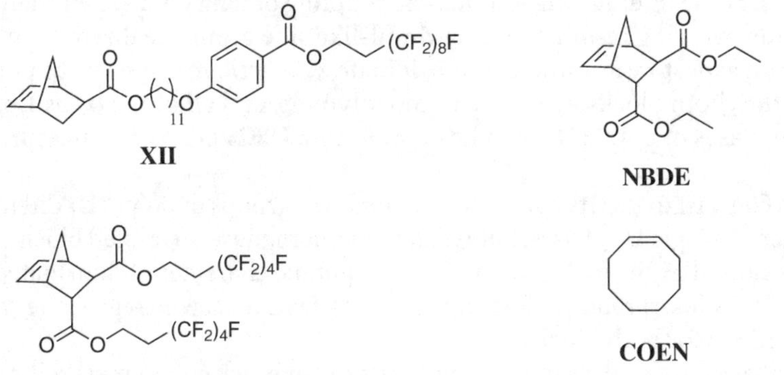

Fig. 11 Monomers used for the preparation of partially perfluorinated block copolymers

Table 9 Physico-chemical properties for the homopolymers poly-XII, poly-XIII, poly-NBDE, and poly-COEN

Polymer	Initiator	M_n	PDI	Phase transitions (°C)[a]
poly-XII	5	87,000	1.38	g 28 s 97 i
poly-XIII	5	N.D.	N.D.	g 16 i
poly-NBDE	4	100,000	1.63	g 32 i
poly-COEN	4	80,000	1.63	g –80 k 65 i

[a] Phase transitions are given for the second heating run and first cooling run, respectively (DSC).

Table 10 Physico-chemical properties of AB-diblock copolymers

Copolymer	Initiator	M_n	PDI	A:B	Block A	Phase transitions (°C)[a]				
						T_g	T_i	Block B	T_g	T_i
poly-XII-b-COEN	5	132,000	1.48	1:8	XII	30	100	COEN	−78	59
poly-XIII-b-NBDE	5	157,000	1.58	1:1.5	XIII	9	–	NBEN	31	–
poly-XIII-b-COEN	5	102,000	1.57	1:6	XIII	11	–	COEN	−71	58

[a] Phase transitions are given for the second heating run and first cooling run, respectively (DSC).

identified sequence of mesophases of poly-**XII** was smectic I_2 – smectic F_2 – smectic C_2 with rising temperature, as identified by Wide Angle X-ray Diffraction (WAXD). No transition temperatures were given for this sequence in the article. These findings suggest that the coupling of long fluorinated chains with an aromatic ring resulted in a rigid rod-like side group (realized in poly-**XII**) with an aspect ratio sufficient to originate a smectic mesophase. In poly-**XII**-b-**COEN**, both blocks acted like homopolymers, and all of the transitions and mesophases of poly-**XII** could be obtained via DSC and WAXD measurements too.

In contrast to that, the short fluorinated side groups of poly-**XIII** did not lead to thermotropic LC phases, but resulted in microphase-separated block copolymers caused by the hydrophobic and oleophobic character of fluorinated polymers. This interpretation was supported by DSC data, which gave the thermal transitions for both blocks.

Koltzenburg et al. report the synthesis of AB block copolymers of acetylene and norbornene derivatives bearing mesogenic moieties [68]. A norbornene

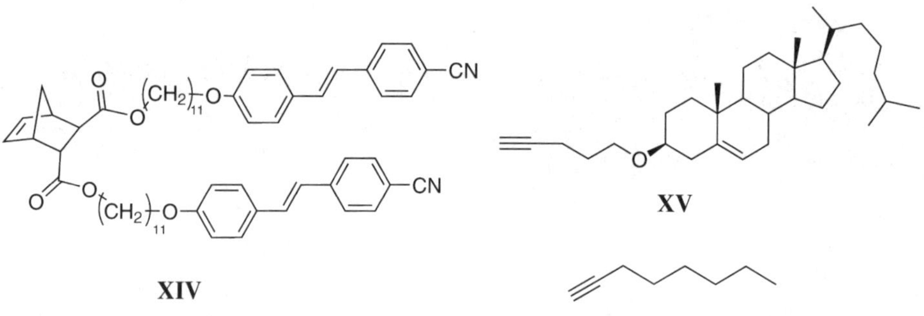

Fig. 12 Monomers used for the copolymerization of norbornene derivatives with alkynes

derivative with cyanostilbenyl side groups **XIV** and an acetylene derivative with cholesteryl group **XV** or 1-octyne were prepared, as shown in Fig. 12. Polymerizations were carried out with initiator **1**. Homopolymers poly-**XIV** and poly-**XV** as well as block-copolymers poly-**XIV**-*b*-**XV** and poly-**XIV**-*b*-(octyne) formed thermotropic mesophases. The acetylene polymer poly-**XV** showed a smectic A phase, while the poly(norbornene) poly-**XIV** was nematic. The block copolymer poly-**XIV**-*b*-**XV** showed microphase separation, retaining the homopolymers' mesophases.

2.3 Laterally-Attached Mesogens

In SCLCPs with laterally-attached mesogens, the spacer is linked to the middle of the mesogenic unit. Therefore, the symmetry of mesogens substituted with two long alkyl chains – a very common motive in mesogens – is less disturbed than by terminal attachment. In this kind of SCLCP, the nematic mesophase is favored over other phases. Due to the large size of the mesogenic unit compared to the repeating unit of the polymer backbone, the nematic orientation of the

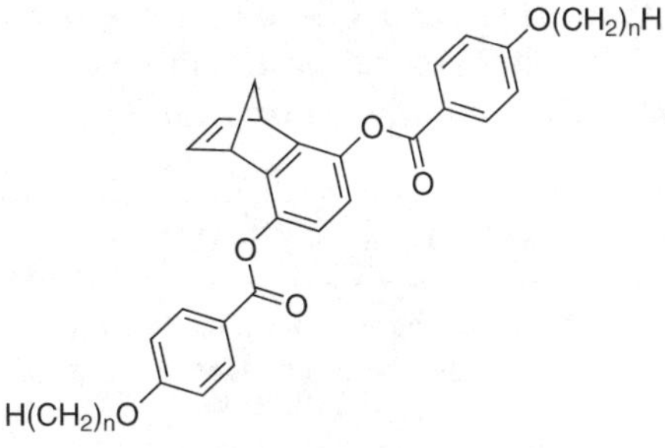

XVI-n, n = 1-9

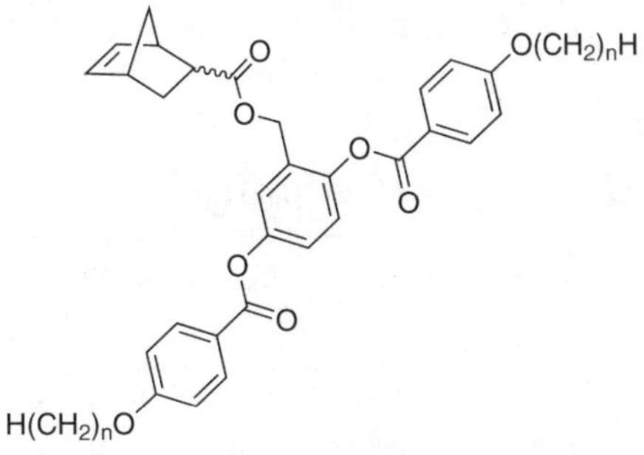

XVII-n, n = 1-6

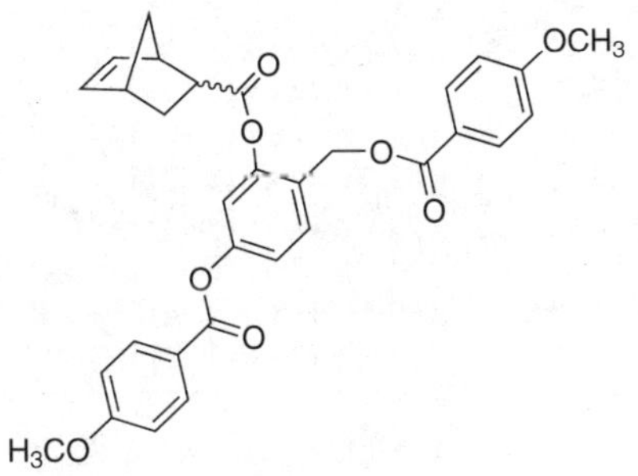

XVIII-1

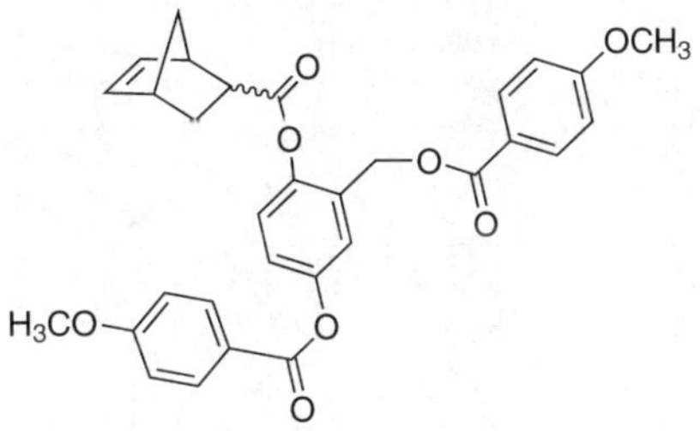

XIX-1

Fig. 13 Structural representations of monomers **XVI–XIX**

mesogen forces the polymer backbone into an extended helical structure, in which the mesogens wrap around the backbone and form a cylindrical jacket. Smectic phases can be induced by long spacer groups and/or the introduction of immiscible moieties like fluorocarbons into the mesogenic unit [40].

Norbornene derivatives with laterally attached 2,5-bis[(4'-n-alkoxybenzoyl)oxy]phenyl mesogens were prepared by Pugh et al. (Fig. 13). [40, 69]. Even the fusion of the middle benzene ring of the mesogens to the norbornene (XVI-n, n=1–9), as well as separation by a COOCH$_2$-spacer (monomers XVII-n, n=1–6), led to substances with a nematic liquid crystalline phase. Monomers XVIII-1 and XIX-1 were constitutional stereoisomers of XVII-1. The introduction of the flexible CH$_2$ unit into the mesogenic units altered the rigidity of the mesogenic unit and no LC phases were observed.

Polymerization of monomers XVI-n and XVII-n with initiator 4 led to SCLCPs with nematic mesophases. Polymers poly-XVI-n had high T_g values (in the range of 160–200 °C) and formed nematic phases. The polymers did not show any isotropization temperature, but decomposed to a certain degree, as illustrated by the darkening of the samples during polarized optical microscopy (POM) observations. The lateral arrangement forces the polymer to adopt a nematic mesophase. The dependence of the phase transition temperatures on the polymerization degree (DP) was investigated by varying the polymerization degree of poly-(XVII-1) from 5 to 100 (Table 11). Both T_g and T_i became independent of molecular weight at approximately 25 repeating units – an effect already seen in terminally-attached mesogens [38]. The isotropization tem-

Table 11 SEC and DSC data for poly-(XVII-n), poly-(XVIII-1), and poly-(XIX-1)

Monomer	M_0/I_0	M_n	PDI	DP	Phase transitions (°C) and enthalpy changes (kJ mru^{-1}, in parentheses)[a]
XVII-1	3.8	2670	1.16	5.1	g 79.2 n 131.4 (1.80) i
XVII-1	9.7	4320	1.20	8.2	g 89.6 n 145.6 (2.16) i
XVII-1	19.5	7707	1.18	14.6	g 93.9 n 156.4 (2.88) i
XVII-1	49.6	20,773[b]	1.13[b]	39.3[b]	g 96.9 n 162.6 (2.76) i
XVII-1	30.7	24,051	1.13	45.5	g 97.7 n 163.5 (2.76) i
XVII-1	68.6	52,631[b]	1.16[b]	99.6[b]	g 97.1 n 162.5 (2.63) i
XVII-2	39.0	17,990[b]	1.19[b]	32.3[b]	g 92.5 n 172.2 (3.96) i
XVII-3	45.3	25,826[b]	1.24[b]	44.2[b]	g 83.0 n 140.2 (2.88) i
XVII-4	4.7	5174	1.16	8.4	g 62.8 n 122.8 (3.04) i
XVII-4	39.2	14,316	2.17	23.4	g 73.4 n 138.5 (3.57) i
XVII-5	50.8	28,748[b]	1.18[b]	44.8[b]	g 60.3 n 123.2 (3.14) i
XVII-6	52.4	44,082[b]	1.24[b]	65.9[b]	g 55.6 n 126.1 (3.72) i
XVIII-1	60.7	21,235	1.14	40.2	g 78.9 i
XIX-1	158.6	89,008	1.20	168.4	g 84.1 i

[a] Phase transitions are given for the second heating run (DSC).
[b] Fractionated polymer.

peratures of poly-(**XVII-n**) showed an odd-even effect; spacers with odd numbers of methylene units had lower isotropization temperatures.

In order to induce smectic layering in SCLCPs with laterally attached mesogens, Pugh et al. suggested three different approaches [32, 70]. In the first one, mesogens, which exhibit the smectic C – nematic – isotropic phase sequence, as in 1,4-bis[(3′fluoro-4′-*n*-alkoxyphenyl)-ethenyl]benzenes, are linked to norbornenes; **XX-n**, **XXI-n** in Fig. 14. In the second approach, smectic layering was attempted by replacing the terminating alkoxy-groups of **XVII-n** by a partially perfluorinated group, as realized in the precursor **XXIII-m-n** (see Fig. 14). The immiscibility of the fluorocarbon and hydrocarbon segments can induce smectic layering. This effect had been observed for oligomers consisting of diblocks and triblocks of composition $H(CH_2)_n(CF_2)_mF$ or $F(CF_2)_m(CH_2)_n(CF_2)_mF$ ($m \geq 6$ and $4 \leq n \leq 14$). These oligomers underwent a crystalline-smectic phase transition prior to melting [71]. The third was

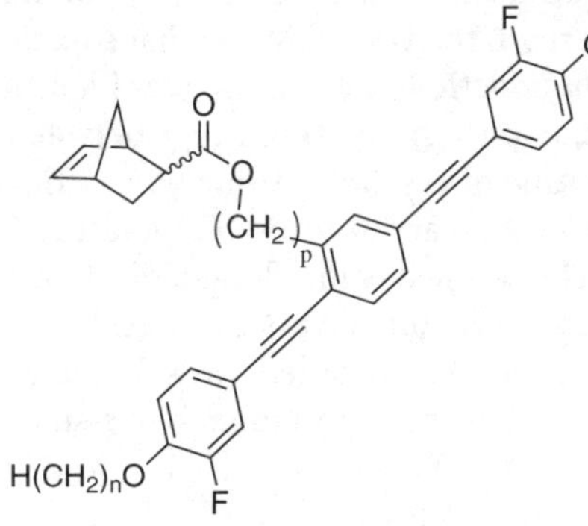

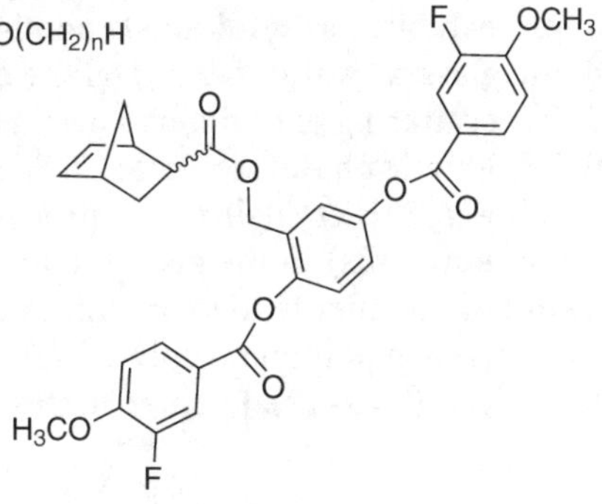

p=1: **XX-n**
p=11: **XXI-n**
n = 1-12

XXII

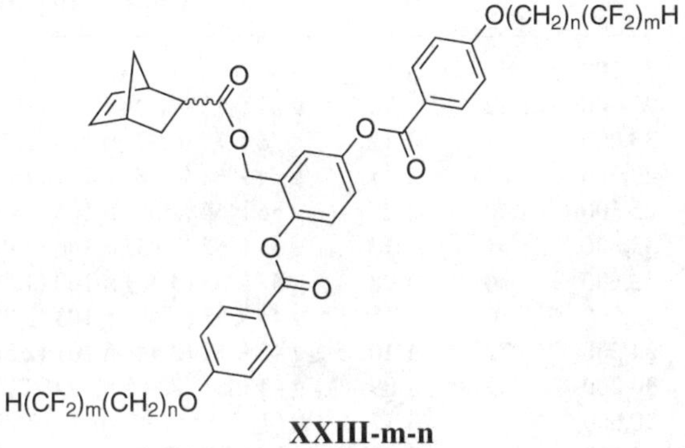

XXIII-m-n
n = 3-6, 8
m = 2-4, 6-8

Fig. 14 Monomers with laterally-attached mesogens featuring fluoro substituents

copolymerization of one monomer containing an electron-rich mesogen with a second monomer containing an electron-poor mesogen. The electron-donor-acceptor interactions of adjacent mesogenic units were expected to form layered structures, facilitating smectic mesophases.

Following the idea of the first approach, monomers with the 1,4-bis[(3′-fluoro-4′-n-alkoxyphenyl)-ethynyl]benzene mesogens were laterally linked via a CH_2-spacer to norbornene, **XX-n** (n=1–12). The low molecular mesogens with $n≥6$ showed a k-S_E-S_C-n-i phase sequence, whereas the norbornene monomers **XX-n** exhibited only a monotropic or enantiotropic nematic mesophase, respectively [72].

Polymerizations of **XX-n** were carried out with Schrock-type initiator **4** in a monomer:initiator ratio of about 50, to be independent of polymer chain length, leading to well-defined polymers with low PDIs from 1.08–1.29, as shown in Table 12.

All polymers, even those with relative short spacers (n=1–3), showed an enantiotropic nematic LC phase. After annealing for several days, polymers with $n≥3$ exhibited an additional endotherm in the DSC diagram that coincided with the glass transition. A crystalline or smectic E mesophase was identified by X-ray scattering experiments over a narrow temperature range between T_g and the nematic phase; see Table 12. The time of crystallization decreased with increasing n. This crystallization process took place at room temperature (for up to nine months) in the glassy state. This indicates that long side chains in this kind of polymer tend to organize very slowly into layers, an effect not seen in the reference polymers, poly-(**XVII-n**). In more detailed WAXD studies of poly-(**XX-n**) (n=9–12), Pugh et al. showed that these polymers had a smectic

Table 12 Physico-chemical data for poly-(**XX-n**)

Monomer	M_0/I_0	M_n	DP	PDI	Phase transitions (°C) and enthalpy changes (kJ mru^{-1}, in parentheses)[a]
XX-1	44	33,300	63	1.23	g 90 n 122 (1.54) i
XX-2	51	23,100	42	1.14	g 87 n 139 (2.48) i
XX-3	46	33,000	57	1.12	g 76 k 74 (0.87) n 115 (1.58) i
XX-4	50	40,700	67	1.11	g 64 k 68 (1.58) n 115 (1.74) i
XX-5	50	25,500	40	1.20	g 56 k 64 (2,09) n 105 (1.49) i
XX-6	50	33,600	51	1.14	g 53 k 62 (2.95) n 106 (2.03) i
XX-7	87	52,600	76	1.08	g 47 k 55 (3.36) n 100 (1.81) i
XX-8	46	42,900	60	1.15	g 45 k 55 (3.48) n 103 (2.22) i
XX-9[b]	52	54,200	72	1.10	g 43 k 52 (2.44) n 101 (2.50) i [b]
XX-10[b]	46	30,200	39	1.09	g 44 k 51 (2.90) n 99 (2.77) i [b]
XX-11[b]	47	38,600	48	1.17	g (41 k 51 (4.42) n 99 (2.75) i [b]
XX-12[b]	47	45,500	55	1.29	g 36 k 47 (3.69) n 93 (2.52) i [b]

[a] Phase transitions are given for the second heating run (DSC).
[b] See also Table 13.

Table 13 Phase transitions (°C) and enthalpy changes (kJ mol^{-1}, in parentheses) for poly-(XX–n)

n	T_{Sc-N}	T_{N-I}
XX–9	97 (0.9)	101 (1.9)
XX–10	96 (1.0)	99 (2.0)
XX–11	95 (1.3)	98 (2.2)
XX–12	87 (0.8)	91 (2.3)

C – nematic mesophase sequence, and in the case of $n=12$ the tilt angle of the smectic C phases decreased with temperature, resulting in the formation of an additional smectic A phase [73]. The phase transitions of these polymers are given in Table 13. WAXD studies of poly-(XX–n), $n=2-7$ confirmed that these polymers had a nematic phase. Some additional structural features in the X-ray pattern were interpreted as smectic C fluctuations. Poly-(XX–8), however, showed a smectic C mesophase similar to those of poly-(XX–n), $n=9-12$ [74].

Following the same concept, the short spacer p=1 was exchanged with a long alkyl spacer with 11 methylene units, and monomers XXI–n, with $n=1-12$, were synthesized (Fig. 14) [75]. Increasing the spacer length led to a lower mesogen density in the final polymer. All monomers, except XXI–1 and XXI–3, showed a nematic mesophase. Polymerization of XXI–n was carried out using initiator 4 with a monomer to initiator ratio of about 50. The molecular weight data and the thermal transition are listed in Table 14. The synthesized polymers had relatively low PDIs (1.11–1.37), and their thermal transition temperatures were independent of their molecular weights when the degree of polymerization was >25, as shown by a series of different sizes of poly-(XXI–9) in Table 14. The goal, to synthesize SCLCPs with smectic mesophases, could not be achieved by this route. All polymers showed an enantiotropic nematic phase, except poly-(XXI–1), which did not show any LC phase. In contrast to poly-(XX–n) with the short C_1-spacer, these polymers had a reduced tendency to crystallize. This indicated that it was the poly(norbornene) backbone that crystallizes rather than the side chains.

The synthesis and polymerization of monomer XXII with initiator 4 was mentioned by Pugh et al. [32]. Poly-XXII exhibited a nematic mesophase between 103–128 °C.

Following the second approach, Pugh and co-workers synthesized a series of norbornene derivatives with laterally attached mesogens of type XVII–n, where the terminating alkoxy group was replaced by a partially perfluorinated endgroup, XXIII–n–m, with $n=3-6, 8$ and $m=2-4$ and $6-8$ (Fig. 14) [76-78]. This route used the immiscibility of hydrocarbons with fluorocarbon segments, which was expected to force the polymer to adopt a layered structure, favoring a smectic mesophase. In a first work, mesogens with relatively long perfluorinated endgroups, $m=6-8$, were investigated, based on the observation that

Table 14 SEC and DSC data for poly-(XXI–n)

Monomer	[4]:[XXI–n]	PDI	M_n	DP	Phase transitions (°C) and enthalpy changes (kJ mru^{-1}, in parentheses)[a]
XXI–1	1:50	1.17	44,200	66	g 24 k 29 (0.47)
XXI–2	1:50	1.29	36,300	51	g 25 k 31 (1.58) n 48 (1.41) i
XXI–3	1:42	1.19	25,800	36	g 29 k 35 (2.38) n 47 (0.88) i
XXI–4	1:50	1.13	36,700	49	g 15 n 45 (1.27) i
XXI–5	1:50	1.14	53,400	68	g 8 n 40 (1.38) i
XXI–6	1:50	1.21	35,400	44	g 3 n 48 (2.08) i
XXI–7	1:50	1.18	42,300	50	g 0 n 44 (1.86) i
XXI–8	1:50	1.16	41,700	49	g –2 n 51 (2.69) i
XXI–9	1:5	1.20	12,900	14	g –4 k 37 (15.7) n 41 (1.92) i
XXI–9	1:20	1.11	15,400	17	g –3 k 40 (14.2) n 46 (1.47) i
XXI–9	1:35	1.26	24,100	27	g –3 k 38 (1.53) n 47 (1.43) i
XXI–9	1:55	1.19	52,000	58	g –3 k 39 (16.5) n 48 (2.32) i
XXI–9	1:80	1.37	80,800	91	g –3 k 37 (6.29) n 49 (2.54) i
XXI–9	1:100	1.37	140,000	158	g –3 k 38 (0.76) n 50 (2.19) i
XXI–10	1:50	1.13	44,200	48	g –4 n 52 (3.10) i
XXI–11	1:42	1.15	35,100	37	g –5 k 39 (11.1) n 50 (3.16) i
XXI–12	1:50	1.20	44,300	45	g –3 n 53 (3.98) i

[a] Phase transitions are given for the second heating run (DSC).

saturated molecules containing fluorocarbon segments with at least six CF_2 groups organize into layers [76].

All of the monomers **XXIII–n–m** with n=4–6, 8 and m=6–8 showed a monotropic smectic C and an enantiotropic smectic A phase, whereas the pure hydrocarbon analogues formed only a monotropic nematic phase [40]. Polymerization was carried out with initiator 4 in a monomer to initiator ratio of about 50:1. The final purified polymers had degrees of polymerization ranging from 8 to 52, and rather high PDIs of between 1.18 and 2.06, due to immiscibility and bimodal polymerization problems. However, it was possible to prove that introducing fluorocarbon moieties into the mesogenic unit leads to formation of a smectic C phase over a broad temperature range. A very narrow smectic A phase was also observed shortly below T_i. The second part of Table 15 summarizes the physico-chemical data for poly-(**XXIII–n–m**). Preliminary results from small angle X-ray scattering (SAXS) experiments confirmed the smectic A phase. The segregation effect of the fluorocarbon segments turned out to be so strong that the polymer backbone's influence was minimal, with the polymers' transition temperatures being similar to those of the model compounds 2,5-bis-[4′-(n-(perfluoroalkyl)-alkoxy)-benzoyl)oxy)-toluenes. In general, the melting points of the model compounds were simply replaced by a glass transition and the isotropization temperatures were about 20 °C higher in the polymers than in the model compounds.

Table 15 Molecular weight and thermal transitions for poly-(**XXIII-m-n**)

n	m	[XXIII]:[4]	PDI	M_n	DP	Phase transitions (°C) and enthalpy changes (kJ mru^{-1}, in parentheses)[a]	Reference
3	2	50	1.43	32,400	40	g 76 i	[78]
4	2	54	1.47	75,900	89	g 75 n 99 (2.00) i	
5	2	50	1.34	31,200	36	g 65 n 94 (0.77) i	
6	2	52	1.40	22,600	25	g 65 n 102 (1.30) i	
8	2	45	1.49	30,300	31	g 61 n 104 (1.21) i	
3	3	49	1.36	99,800	108	g 81 n 106 (0.86) i	
4	3	52	1.16	13,800	15	g 85 n 108 (0.83) i	
5	3	38	1.28	43,700	45	g 75 n 120 (0.80) i	
6	3	49	1.32	43,600	43	g 90 n 124 (1.18) i	
8	3	51	1.25	27,500	26	g 61 s$_A$ 139 (3.18) i	
3	4	52	1.34	46,100	45	g 96 n 137 (0.710) i	
4	4	50	1.45	60,600	58	g 91 n 144 (1.18) i	
5	4	50	1.57	74,400	69	g 81 s$_A$ 164 (2.59) i	
6	4	53	1.45	43,700	40	g 77 s$_A$ 171 (3.06) i	
8	4	50	1.39	59,900	80	g 81 s$_A$ 169 (4.00) i	
4	6	50	1.39	55,800	52	g 106 s$_C$ 227 (6.97) s$_A$ 234 (4.09)	[76]
5	6	44	1.32	43,100	35	g 96 s$_C$ 228 s$_A$ 231 (4.42) i	
6	6	50	1.39	25,900	19	g 90 s$_C$ 216 s$_A$ 223 (4.36) i	
8	6	50	1.49	52,700	39	g 77 s$_C$ 213 s$_A$ 216 (4.22) i	
4	7	50	1.45	18,400	14	g 90 s$_C$ 242 (3.21) s$_A$ 251 (0.52) i	
5	7	50	1.62	22,600	16	g 96 s$_C$ 239 s$_A$ 248 (4.35) i	
6	7	48	1.28	19,700	14	g 93 s$_C$ 230[b] s$_A$ 236 (3.56) i	
8	7	52	1.22	39,500	27	g 97 s$_C$ 228 s$_A$ 232 (3.45) i	
4	8	51	1.18	12,700	9	g 93 s$_C$ 251 s$_A$ 264 (3.87) i	
5	8	49	1.48	12,000	8	g 93 s$_C$ 258 s$_A$ 262 (3.81) i	
6	8	49	2.06	21,300	14	g 98 s$_C$ 250 s$_A$ 261 (3.78) i	
8	8	50	1.37	25,900	17	g 98 s$_C$ 231 s$_A$ 234 (0.69) i	

[a] Phase transitions are given for the second heating run (DSC).

In a subsequent paper, Pugh and co-workers investigated the LC properties of **XXIII-n-m** with short fluorocarbon endgroups, m=2–4 [78]. Monomers with the shortest fluorocarbon segment, m=2, did not show any mesophase when the hydrocarbon segment was short (n=3–5). However, longer hydrocarbon spacers showed monotropic (n=6) or enantiotropic (n=8) nematic mesophases. Monomers with m=3 were on the borderline for smectic induction. Whereas monomers with n=3 and 4 did not exhibit any mesophase, monomers with n=5, 6 and 8 as well as all monomers with m=4 exhibited an enantiotropic smectic A phase. Polymerization of these monomers was carried out with initiator **4** in a monomer to initiator ratio of about 50. The polymerization was less problematical than for monomers with longer perfluorinated segments, and polymers with low PDIs of between 1.16 and 1.57 were obtained.

Molecular weights and molecular weight distributions, together with the thermal transitions of poly-**XXIII-m-n** are summarized in Table 15. From these data it can be concluded that inducing smectic layering requires a minimum of eight methylene units and three difluoromethylene units or three methylene units and four difluoromethylene units. All other polymers with $m=3,4$ exhibited nematic mesophases. Polymers with $m=2$ displayed only nematic mesophases, or in the case of $n=3$ no mesophase at all. Because of the thermal transition data, the authors concluded that the polymers were not microphase-separated.

Spurred on by discussions on whether the smectic layers of the fluorocarbon-substituted polymers **XXIII-m-n** are induced by the immiscibility between hydrocarbon and fluorocarbon segments or by the fact that fluorocarbon segments form rod-like mesogenic units by themselves, Pugh et al. synthesized a series of siloxane-terminated monomers **XXIV-m-n**, with $m=1,2$ and $n=4-8$ [79]. In these monomers, the rigid fluorocarbon segment was replaced by short polydimethylsiloxane (PDMS) segments that are also immiscible with hydrocarbons but are very flexible (Fig. 15).

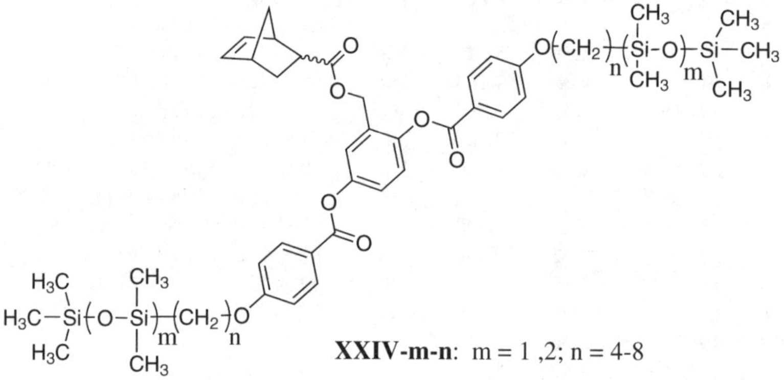

XXIV-m-n: m = 1,2; n = 4-8

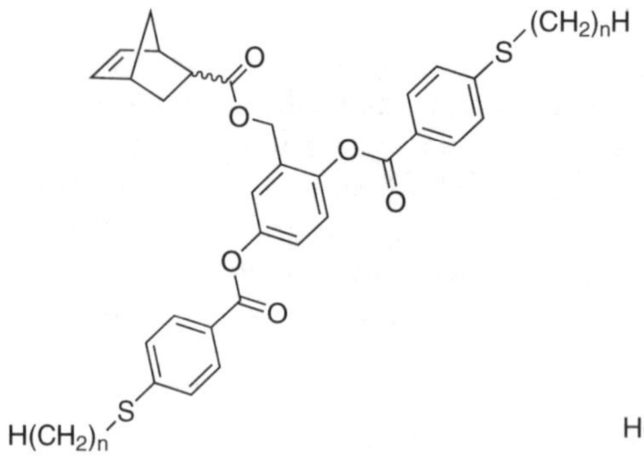

XXV-n: n = 1-12

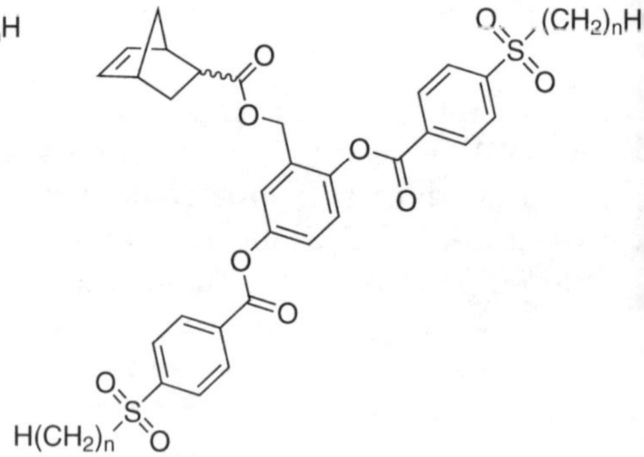

XXVI-n: n = 1-12

Fig. 15 Monomers designed for inducing smectic layering following concepts 2 and 3

Table 16 Physico-chemical data for poly-(XXIV–m–n)

n	m	[XXIV]:[4]	PDI	M_n	DP	Phase transitions (°C) and enthalpy changes (kJ mru^{-1}, in parentheses)[a]
4	1	48	1.57	46,900	52	g 26 s$_C$ 38 (2.03) i
5	1	41	1.51	42,400	45	g 48 s$_C$ 57 (4.70) i
6	1	45	1.51	50,800	53	g 46 s$_C$ 56 (5.55) i
8	1	49	1.55	70,400	69	g 51 k 60 (2.50) s$_C$ 72 (3.37) i
4	2	48	1.37	41,200	39	g 17 s$_C$ 23 (2.11) i
5	2	50	1.26	39,900	37	g 44 s$_C$ 54 (4.99) i
6	2	50	1.87	70,500	63	g 46 s$_C$ 55 (4.10) i
8	2	47	1.65	10,600	91	g 50 k 61 (3.94) s$_C$ 72 (2.47) i

[a] Phase transitions are given for the second heating run (DSC).

The monomers **XXIV–m–n** were crystalline solids and only the monomers with the shortest carbon-segments ($n=4$) showed an enantiotropic smectic C mesophase. Monomers with higher carbon chains did not exhibit a LC phase.

The monomers **XXIV–m–n** were polymerized with initiator 4 in a monomer to initiator ratio of about 50. Table 16 summarizes the physico-chemical data for these polymers. Poly-(**XXIV–m–n**) exhibited smectic C mesophases, but these phases were only observed over a small temperature range and isotropization occurred shortly after the glass transition. It was not possible to identify the mesophase by polarized optical microscopy, but the smectic phase was confirmed by X-ray scattering experiments. Transition temperatures increased with the length of the carbon segments (n) but decreased with increasing siloxane segment (m).

Physical data for these polymers supported the idea that induction of smectic layering stems from the immiscibilities of the hydrocarbon and the oligodimethylsiloxane or fluorocarbon segments, and not from the rigid rod-like shape of the fluorocarbon segments.

Following the third concept described above, of using electron-donor-acceptor interactions for inducing smectic mesophases in SCLCPs, Pugh et al. presented the thioether monomers **XXV-n** as electron-donating and the sulfone derivatives **XXVI-n** as electron-accepting entities in a preliminary paper, as shown in Fig. 15 [80].

2.3.1
Block Copolymers

A block copolymer consisting of a SCLCP-block of monomer **XXVII** with a laterally-attached mesogenic unit, and butyl-acrylate, was synthesized using a combination of ROMP and atom transfer radical polymerization (ATRP) (Fig. 16) [81].

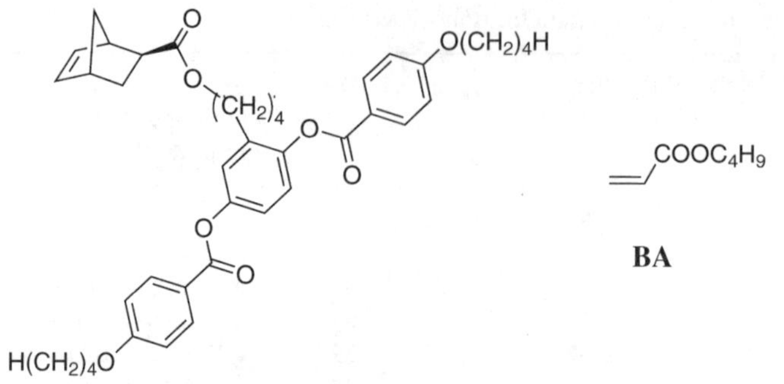

Fig. 16 Monomers used for block copolymerization containing laterally-attached mesogens

ROMP of the SCLCP block was carried out with the "Grubbs"-initiator **6** in a monomer to initiator ratio of about 25. The polymerization was terminated by reaction with 4-(2-bromopropionyloxy)-but-2-enyl 2-bromopropionate, leading to a macroinitiator for ATRP. The radical polymerization was carried out with CuCl, 4,4'-di(*n*-nonyl)-2,2-bipyridine, and butylacrylate (**BA**), giving a poly-**XXVII**-*b*-**BA** copolymer. The homopolymer of poly-**XXVII** had a narrow PDI of 1.06, while the diblock copolymer showed a PDI of 1.32.

The homopolymer showed an enantiotropic nematic mesophase, whereas the diblock copolymer generated microphase-separated lamellae, in which the SCLCP block possessed a nematic-isotropization transition similar to the homopolymer (Table 17). Upon heating, the nematic microphase decreased continuously in the nematic phase from 38.5 nm to 27 nm and showed a constant value of about 26 nm after the nematic-isotropization transition. Therefore, materials in which these block copolymers are macroscopically aligned are expected to show reversible contraction in one dimension, making this polymer system an interesting candidates for an artificial muscle or actuator.

Table 17 Molecular weights and thermal transitions of poly-**XXVII** and poly-**XXVII**-*b*-**BA**

Polymer	M_n	PDI	Phase transitions (°C) and enthalpy changes ($J\,g^{-1}$, in parentheses)[a]
poly-**XXVII**	22,900	1.06[a]	k 60.9 (8.48) n 102.5 (2.24) i
poly-**XXVII**-*b*-**BA**	36,100	1.28[b]	g –46 (poly-BA) k 55.8 (3.15) n 103.3 (1.97) i

[a] Phase transitions are given for the second heating run (DSC).

2.4
Discotic Mesogens

In contrast to calamitic mesogens, discotic liquid crystals are built from disk-like molecules that can arrange into different structures, such as the discotic nematic mesophase, the discotic columnar mesophase, or the discotic hexagonal mesophase.

Discotic liquid crystalline materials, like triphenylenes, show high charge carrier mobilities within their highly ordered mesophases. This enhanced photoconductivity has been attributed to long-range ordering along the columns in the discotic hexagonal and helical phase. In combination with polymers, discotic phases can be arranged in a mechanically stable manner for use in industrial applications. Possible architectures for discotic liquid crystalline polymers are pictured in Fig. 17.

Discotic SCLCPs were synthesized by the group of Grubbs [82]. For this purpose, norbornenes **XXVIII–n** (n=5, 10) and cyclobutenes **XXIX–n** (n=5, 10) with alkoxy-substituted triphenylenes as mesogenic units (see Fig. 18) were prepared. Polymerization was carried out with initiator **6**. The resulting polymers had a narrow PDI between 1.09 and 1.17. Physico-chemical data for poly-**XXVIII** and poly-**XXIX** are listed in Table 18.

DSC measurements and X-ray-scattering experiments demonstrated that polymers with decyloxy-substituted triphenylenes exhibit discotic hexagonal

Fig. 17 Discotic polymer architectures. A: MCLCP; B: SCLCP; C: discotic network

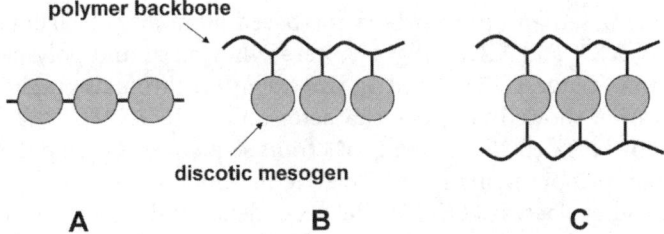

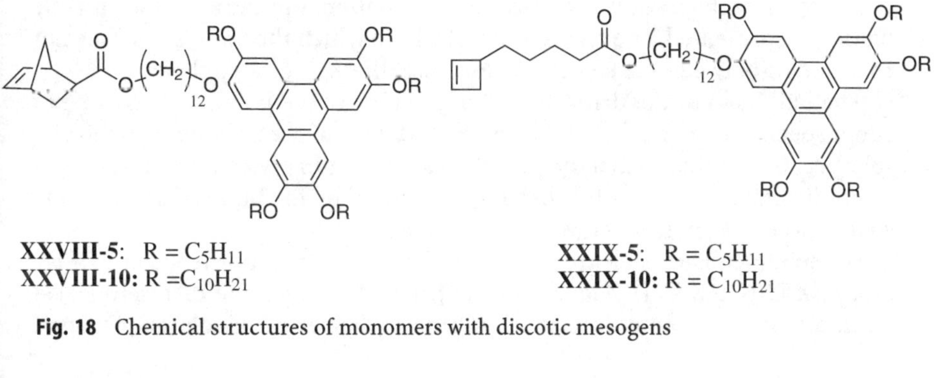

XXVIII-5: R = C_5H_{11}
XXVIII-10: R = $C_{10}H_{21}$

XXIX-5: R = C_5H_{11}
XXIX-10: R = $C_{10}H_{21}$

Fig. 18 Chemical structures of monomers with discotic mesogens

Table 18 Physico-chemical data for poly-(XXVIII-n) and poly-(XXIX-n)

Polymer	M_n	PDI	Phase transitions (°C)
poly-(XXVIII-5)	46,500	1.09	g −3 D_{hd1} 36 D_{hd2} 42 i
poly-(XXVIII-10)	48,500	1.17	g −4 i
poly-(XXIX-5)	33,000	1.11	g −17 D_{hd1} 37 D_{hd2} 45 i
poly-(XXIX-10)	157,000	1.19	g −12 i
poly-(XXIX-5)a	50,000	1.11	g −18 D_{hd1} 34 D_{hd2} 43 i
poly-(XXIX-10)a	125,000	1.32	g −17 i

(D_h) mesophases up to 40 °C, but polymers containing the pentyloxy-substituted triphenylenes do not show any LC phase at all.

Based on previous work which showed that backbone rigidity correlates with mesophase behavior for SCLCPs bearing calamitic mesogens [64], the double bonds of the main chain in polybutadienes were hydrogenated, leading to pure saturated alkane main chains, poly-(**XXIX-5**)a and poly-(**XXIX-10**)a in Table 18. But in contrast to the studies of calamitic SCLCPs, no dependency of backbone rigidity on mesophase behavior could be established.

2.5
Dendritic Side Chains

Norbornene-based and oxa-norbornene-based monomers bearing dendritic side chains, **XXX** and **XXXI** (Fig. 19), were synthesized and polymerized via ROMP with initiator **6** [83]. Based on size exclusion chromatography data, the polymerization shows living-like character up to DP=70. ^{1}H- and ^{13}C-NMR-spectroscopy revealed 35% *cis* and 65% *trans* sequences. These polymers displayed enantiotropic nematic and smectic mesophases, except for DP=5. In contrast to other classes of SCLCPs, the dependence of the DP on the transition temperature of the polymer was very weak. Glass transition and isotropization temperatures became independent of molecular weight above a degree of polymerization of about 10.

X-ray characterization revealed non-conventional packing in the smectic phase. It represented a novel class of SCLCPs in which the main chain was not confined in-between the smectic layers but rather penetrated them.

The bulkiness of dendritic side chains can force the polymer backbone to adopt special geometries [84]. The overall contour of the polymer can then be spherical or rod-like so that the polymers adopt liquid crystalline phases. Even when the driving force of this behavior is caused by the bulkiness of the side chain, these polymers show MCLCP-like behavior.

Different dendritic side groups were linked to 7-oxanorbornene via an ester group (**XXXII–XXXVI**) (Fig. 20) [84–86]. The monomers by themselves also assembled into a supramolecular arrangement of spherical or columnar shapes.

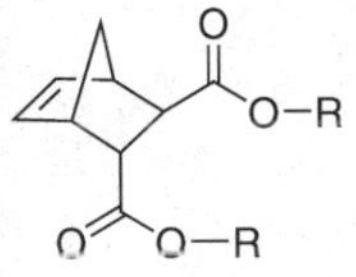

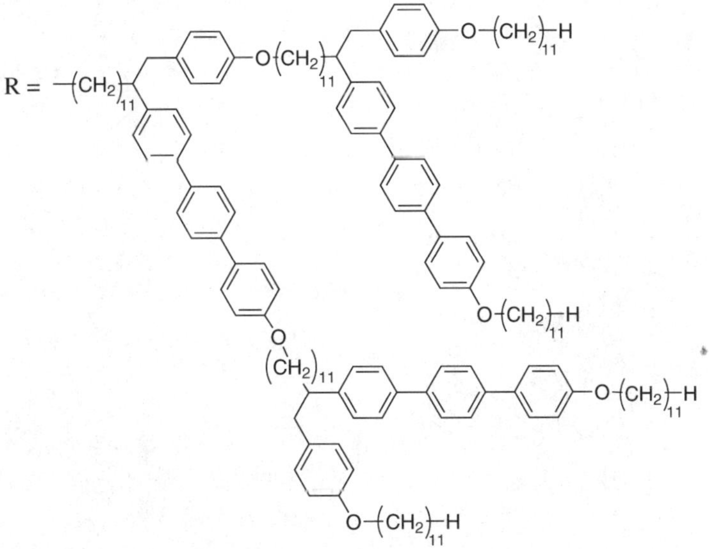

Fig. 19 Chemical structures of monomer **XXX** and **XXXI**

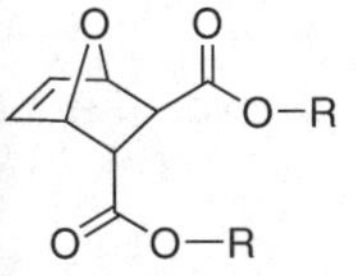

Generally, these pre-self-assembled structural motives are often found in the final polymer materials.

RuCl$_3$·xH$_2$O and RuCl$_2$(Ph$_3$P)$_2$(CHCHCPh$_2$) were used for the polymerization of **XXXII**, leading to polymers with moderate PDIs. Structural investigations revealed a glass – hexagonal columnar – isotropization phase sequence. The more complex monomer **XXXIII** was polymerized using initiator **6**. Monomer **XXXIII**, oligomers, and polymers showed all hexagonal, columnar LC phases. The enthalpy changes and the thermal transition temperatures were divided into three regimes. Up to a DP of 25, the columns were built from self-assembled structures of monomers or oligomers, so enthalpy changes were high and thermal transition temperatures were low. At higher DPs, the columns were built from short polymer entities, and transition temperatures increased up to a plateau for DP≥65.

Polymerization of **XXXIV** was carried out with initiator **6** using monomer to initiator ratios of 20–400, generally resulting in polymers with low PDIs (1.02–1.31). Molecular weight characteristics and thermal transitions of poly-**XXXIV** are listed in Table 19. The polymer chains developed helical structures, forming columns that self-assembled into hexagonal columnar LC phases [84]. The space requirements of the bulky dendritic side groups forced the polymer main chain to adopt a helical structure in the core of a columnar jacket formed by the dendritic side chains. The same structures were found for polymers poly-**XXXV** and poly-**XXXVI**, but oligomers with DPs<25 formed spheres. The change from spherical to columnar polymers was accompanied by an acceleration of the polymerization reaction due reduced sterical hindrance in the columnar arrangement.

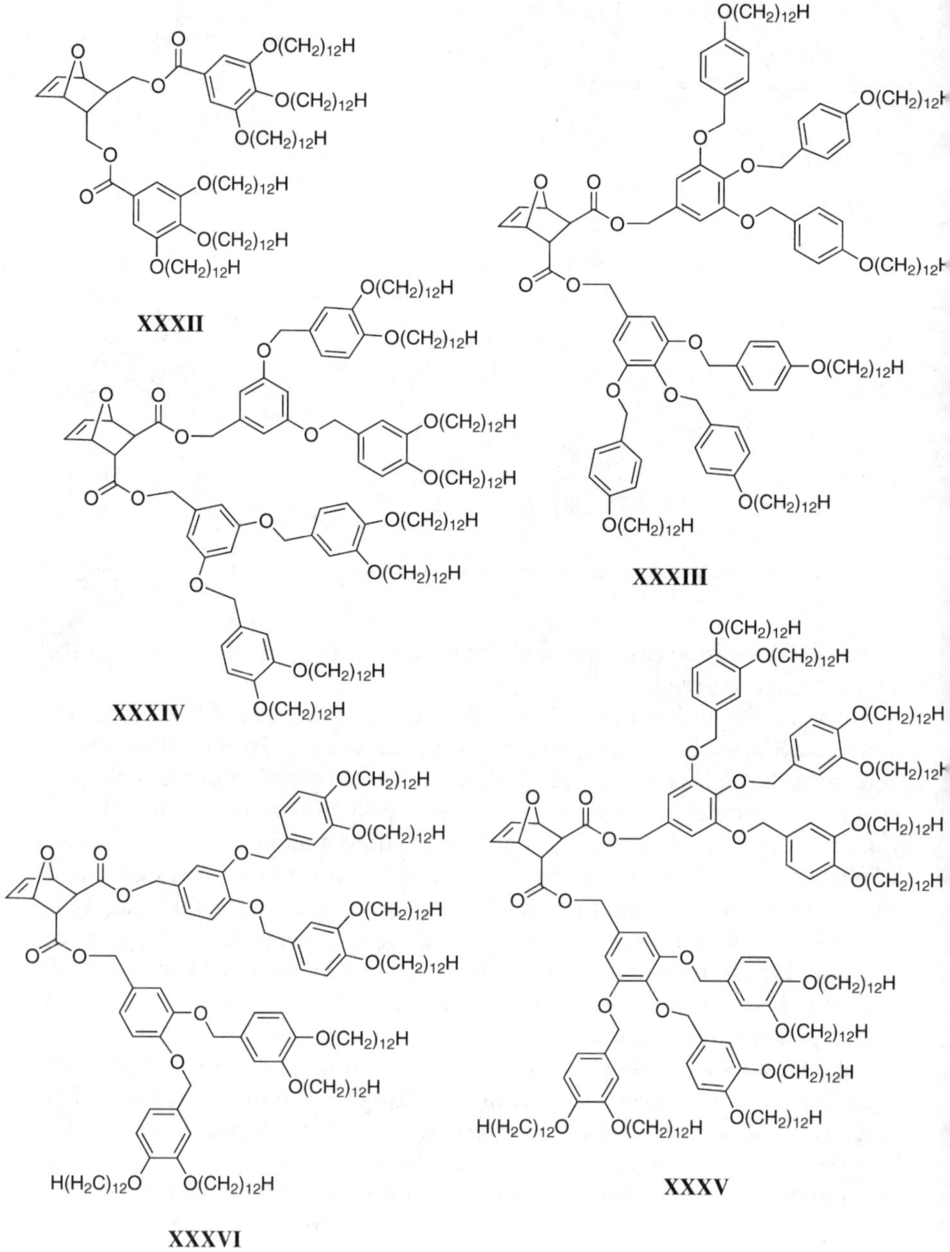

Fig. 20 Monomers **XXXII–XXXVI**, featuring dendritic side chains

Table 19 Physico-chemical data for poly-(**XXXIV**)

[6]:[**XXXIV**]	M_n	PDI	Phase transitions (°C) and enthalpy changes (kcal mru^{-1}, in parentheses)[a]
1:30	45,200	1.04	k −3.7 (3.8) k 95.6 (−0.2) Φ_H 106.2 (0.1) i
1:50	69,200	1.05	k 2.5 (6.3) g 40.6 (−0.2) Φ_H 111.9 (0.1) i
1:75	95,500	1.15	k 2.2 (6.0) g 40.6 Φ_H 116.0 (0.2) i
1:100	158,900	1.06	k 2.1 (4.8) g 41.4 Φ_H 127.9 (0.2) i
1:150	189,800	1.17	k 2.9 (6.0) g 41.6 Φ_H 129.5 (0.2) i
1:200	246,500	1.18	k 2.7 (6.8) g 41.9 Φ_H 132.1 (0.2) i
1:400	511,300	1.13	k 2.6 (5.7) g 41.0 Φ_H 132.9 (0.2) i

[a] Phase transitions are given for the second heating run (DSC).

3
Main Chain Liquid Crystalline Polymers by ADMET und ALTMET

In contrast to ROMP, ADMET offers the possibility of synthesizing both side-chain and main-chain liquid crystalline polymers. The scope and limitations of ADMET are discussed in detail by Wagener et al. in this issue. We herein focus on a few contributions that used step growth polymerization methods to prepare MCLCPs and SCLCPs.

Ferroelectric liquid crystalline monomers (Fig. 21, **XXXVII**) bearing two terminal vinyl groups were polymerized directly from their smectic A* liquid crystal phase using a Grubbs-type initiator.

A degree of polymerization of about ten was determined by end-group analysis via ^{1}H-NMR. Glass transitions were found to occur below room temperature. In each case, the material was smectic C* at 80 °C. The polymers exhibited a broad enantiotropic C* phase range. A uniaxial teflon monolayer gave well-aligned parallel samples which were used to determine the smectic C* tilt angle as a function of temperature. The materials were well-behaved ferroelectric liquid crystals that exhibited large maximum ferroelectric polarization values [87].

Jung-Ii Jin et al. prepared a series of ester-based MCLCPs and realized a combination of MCLCP and SCLCP by copolymerizing the monomers **XXXVIII** and **XXXIX** shown in Fig. 22. The Grubbs catalyst 6 was used for this

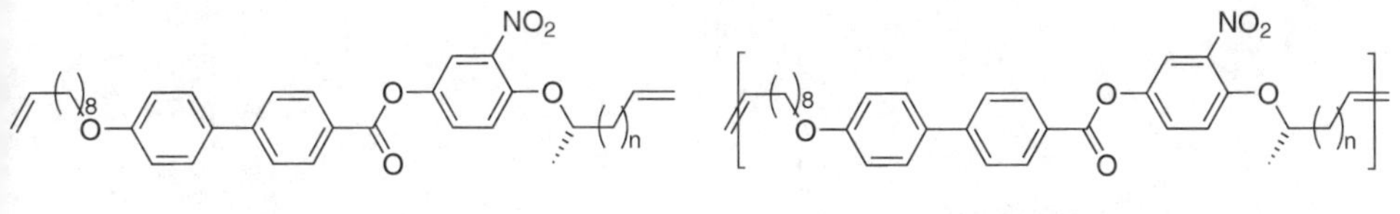

XXXVII n = 2, 3, 4   poly-**XXXVII** n = 2, 3, 4

Fig. 21 Monomers and corresponding polymers prepared by ADMET

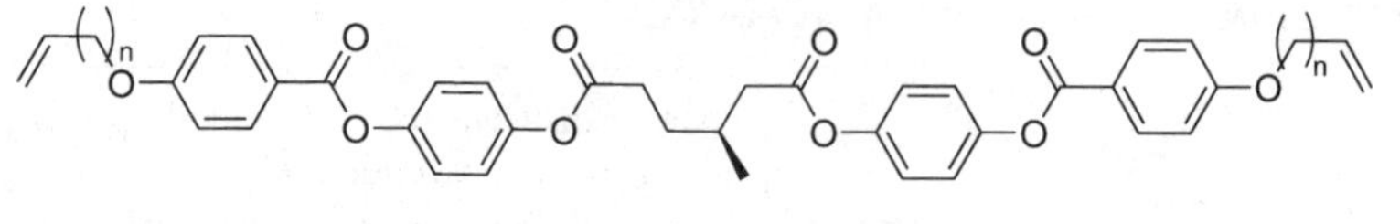

XXXVIII n = 1, 3, 6

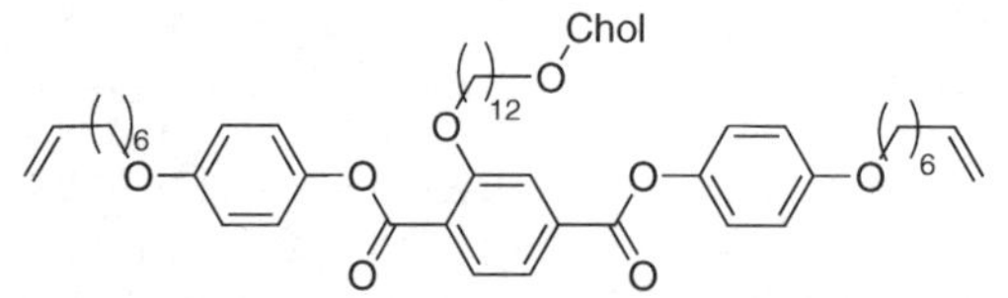

XXXIX Chol = cholesterol

Fig. 22 Chemical structures of the dienes **XXXVIII** and **XXXIX**

purpose. They obtained polymers with degrees of polymerization ranging from 5 to 134. PDIs were determined to be about 2, typical for ADMET polymerizations. Polymers were investigated by DSC and POM. Preliminary WAXD measurements suggested that smectic and nematic mesophases were present. It was outlined that ADMET generally provides a convenient and reliable entry into MCLCPs, and it might be feasible to incorporate functional groups into such polymers that cannot be incorporated by other methods [88].

The same group reported on the synthesis and characterization of all-hydrocarbon MCLCPs, poly-(**XXXX**), and their hydrogenated derivatives poly-(**XXXXI**), based on 4,4′-bis(α-ω-alkenyl)-1,1′-biphenyl derivatives, see Fig. 23. Monomers with different α-ω-alkenyl chain lengths were used to prepare not only the corresponding homopolymers but also statistical copolymers. Crystallinity, thermal transition properties and LC properties were studied.

All polymers showed semicrystalline behavior, as demonstrated by WAXD measurements. DSC thermograms were presented, but LC properties were found to be difficult to determine due to the concomitant thermally-induced crosslinking of the unsaturated polymers upon melting. In some cases smectic B phases were identified by POM. The melting temperatures of the saturated analogues were lower than for the corresponding unsaturated derivatives. Optical textures of the saturated polymers showed lancets in the background, typical of a solid-like smectic phase [89].

poly-**XXXX** n = 1, 3, 6 poly-**XXXXI** n = 1, 3, 6

Fig. 23 All hydrocarbon polymers derived from ADMET

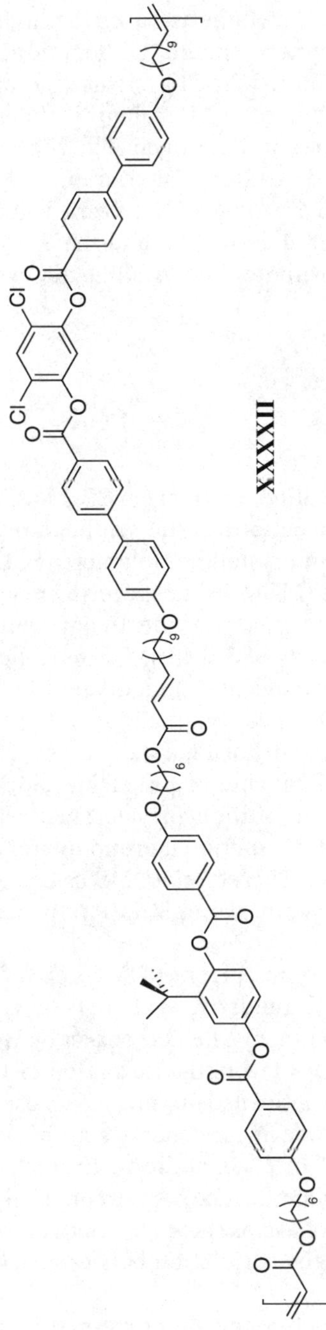

Fig. 24 Repeating unit of ALMET-derived MCLCP

Closing this section, a new metathesis-based method for the preparation of strictly alternating copolymers should be mentioned. Alternating diene metathesis polycondensation (ALTMET) was used to prepare a MCLCP consisting of an alternating architecture of a calamitic (linear rod-like core) and a banana-shaped (bend-shape core) mesogen (Fig. 24).

Characterization revealed a degree of alternation of 99%, an apparent M_W of 43700 g/mol, and a PDI of 1.9. Poly-**XXXXII** formed a nematic phase between the glass transition temperature at 46 °C and the T_i at 120 °C (T_i from DSC 118 °C). Further examples with liquid crystalline diacrylates were disclosed in the publication [90].

4
Conclusion and Outlook

This review of liquid crystalline polymers (LCPs) has highlighted the versatility of olefin metathesis reactions in the synthesis of functional polymers. The chemistry of side chain crystalline polymers (SCLCPs) and main chain liquid crystalline polymers (MCLCPs) has received new impetus from ROMP, ADMET and ALTMET over the last decade. In particular, ROMP has allowed the synthesis of a broad variety of SCLCPs in a controlled way, mostly leading to polymers with narrow molecular weight distributions. Many different mesogen types have been attached terminally as well as laterally to monomers, mainly norbornene and oxanorbornene derivatives. The living character of ROMP makes this method capable of preparing block copolymers just by adding a second monomer. In addition, the latest generation of ruthenium initiators (**8** and **9**) features high functional group tolerance, provides full initiation, and is easy to handle. Therefore, ROMP seems to be one of the most promising approaches for synthesizing SCLCP polymers and copolymers in the future.

Even though the results from current ROMP-derived SCLCPs are not very homogeneous, some general trends are evident. Following the spacer concept of Finkelmann, the introduction of a flexible spacer between the polymer main chain and the mesogenic unit favors the formation of LC phases. The spacer length often has a dramatic influence on the mesophase evolved. The thermal transition temperatures show a linear dependence on molecular weight, reaching a final value above a DP of 25. In addition, the tacticity of the main chain and the Z/E isomerism has an influence not only on the thermal transition temperatures but also the mesophase, as shown by Ungerank [47]. Beside this work, there is little known about the correlation between backbone microstructure and LC phase.

A great number of initiators and monomers are now available, allowing almost perfect control over most of the important parameters of LCPs: main chain stiffness; tacticity; glass transition temperatures; processability from solution or melt; mesogen density along the main chain; combination with different functionalities in statistical and block copolymers. However, this series of easily tunable properties is not yet complete.

Regarding the possible applications of metathesis-derived LCPs, it was shown that the mesophases of SCLCPs can be aligned in magnetic fields, leading to optically transparent materials with high birefringence (Sect. 2.1). These materials are certainly interesting materials for optical applications.

The aligned microphase-separated block copolymer poly-**XXVII**-*b*-**BA** (Sect. 2.3) shows reversible contraction in one dimension – in other words, it acts like an actuator.

Beside classical SCLCPs, attaching dendritic side chains to poly(norbornenes) and poly(7-oxanorbornenes) leads to highly-ordered columnar mesophases (Sect. 2.5). In these polymers, the dendritic side chains force the polymer to adopt a rod-like structure.

MCLCPs prepared by ADMET and ALTMET have recently gained much attention. In particular, ALTMET allows the synthesis of perfectly alternating copolymers, opening new possibilities for the design of materials with specific properties.

References

1. Reinitzer F (1888) Monatsh Chem 9:421
2. Khoo IC (1995) Liquid crystals. Wiley, New York
3. Cifferi A, Krigbaum WR, Meyer RB (eds)(1982) Polymer liquid crystals. Academic, New York
4. Finkelmann H (1987) Angew Chem 99:840
5. McArdle CB (ed)(1989) Side chain liquid crystalline polymers. Blackie, Glasgow, UK
6. Hsu C-S (1997) Prog Polym Sci 22:829
7. Yoon HN, Charbonneau LF, Calundann GW (1992) Adv Mater 4:206
8. Buchmeiser MR (2000) Chem Rev 100:1565
9. Trnka TM, Grubbs RH (2001) Acc Chem Res 34:18
10. Fürstner A (2000) Angew Chem Int Edit 39:3012
11. Frenzel U, Nuyken O (2002) J Polym Sci Pol Chem 40:2895
12. Roy R, Das SJ (2000) Chem Commun 519
13. Jorgensen M, Hadwiger P, Madsen R, Stütz AE, Wrodnigg TM (2000) Curr Org Chem 4:565
14. Maier ME (2000) Angew Chem Int Edit 39:2073
15. Schrock RR (1999) Tetrahedron 55:8141
16. Hoveyda AH, Gillinghan DG, VanVeldhuizen JJ, Kataoka O, Garber SB, Kingsbury JS, Harrity JPA (2004) Org Biomol Chem 2:8
17. Marciniec B, Pietraszuk C (2003) Curr Org Chem 7:691
18. Ivin KJ, Mol JC (1997) Olefin metathesis and metathesis polymerization. Academic, San Diego
19. Fürstner A (ed)(1998) Alkene metathesis in organic synthesis. Springer, Berlin
20. Grubbs RH (ed)(2003) Handbook of metathesis, vols 1–3. Wiley-VCH, Weinheim
21. Kim SH, Lee HJ, Jin SH, Cho HN, Choi SK (1993) Macromolecules 26:846
22. Schrock RR (2003) In: Grubbs RH (ed) Handbook of metathesis, vol 1. Wiley-VCH, Weinheim, p 173

23. Demel S, Riegler S, Wewerka K, Schoefberger W, Slugovc C, Stelzer F (2003) Inorg Chim Acta 345:363
24. Nguyen T-Q, Wu J, Tolbert SH, Schwartz BJ (2001) Adv Mater 13:609
25. Slugovc C, Demel S, Stelzer F (2002) Chem Commun 2572
26. Slugovc C, Riegler S, Hayn G, Saf R, Stelzer F (2003) Macromol Rapid Commun 24:435
27. Love JA, Morgan JP, Trnka TM, Grubbs RH (2002) Angew Chem Int Edit 41:4035
28. Choi T-L, Grubbs RH (2003) Angew Chem Int Edit 42:1473
29. Finkelmann H, Ringsdorf H, Wendorff JH (1978) Macromol Chem 179:273
30. Finkelmann H, Happ M, Portugal M, Ringsdorf H (1978) Macromol Chem 179:2541
31. Pugh C, Kiste AL (1997) Prog Polym Sci 22:601
32. Pugh C, Liu H, Arehart SV, Narayanan R (1995) Macromol Symp 98:293
33. Chen SH, Shi H, Mastrangelo JC, Ou JJ (1996) Prog Polym Sci 21:1211
34. Davidson P (1996) Prog Polym Sci 21:893
35. Walther M, Finkelmann H (1996) Prog Polym Sci 21:951
36. Poser S, Fischer H, Arnold M (1998) Prog Polym Sci 23:1337
37. Komiya Z, Pugh C, Schrock RR (1992) Macromolecules 25:6586
38. Komiya Z, Pugh C, Schrock RR (1992) Macromolecules 25:3609
39. Komiya Z, Schrock RR (1993) Macromolecules 26:1393
40. Pugh C, Schrock RR (1992) Macromolecules 25:6593
41. Gangadhara, Campistron I, Thomas M, Reyx D (1998) J Polym Sci Pol Chem 36:2807
42. Ungerank M, Winkler B, Eder E, Stelzer F (1995) Macromol Chem Phys 196:3623
43. Winkler B, Ungerank M, Stelzer F (1996) Macromol Chem Phys 197:2343
44. Danch A, Lohner K, Ungerank M, Stelzer F (1998) J Therm Anal 54:161
45. Danch A, Kocot A, Ziolo J, Stelzer F (2001) Macromol Chem Phys 202:105
46. Wewerka A, Viertler K, Vlassopoulos D, Stelzer F (2001) Rheol Acta 40:416
47. Ungerank M, Winkler B, Eder E, Stelzer F (1997) Macromol Chem Phys 198:1391
48. Sunaga T, Ivin KJ, Hofmeister GE, Oskam JH, Schrock RR (1994) Macromolecules 27:4043
49. Boamfa MI, Viertler K, Wewerka A, Stelzer F, Christianen PCM, Maan JC (2003) Phys Rev Lett 90:025501/1
50. Boamfa MI, Viertler K, Wewerka A, Stelzer F, Christianen PCM, Maan JC (2002) Mol Cryst Liq Cryst 375:143
51. Fuchs G (2004) PhD Thesis. Graz University of Technology, Graz
52. Alig I, Danch A, Stelzer F (unpublished work)
53. Winkler B, Rehab A, Ungerank M, Stelzer F (1997) Macromol Chem Phys 198:1417
54. Cho I, Jo S-Y (1999) Macromolecules 32:521
55. Cho I (2003) Macromol Symp 195:89
56. Dubois JC, Decobert G, LeBarny P, Esselin S, Friedrich C, Nöel C (1986) Mol Cryst Liq Cryst 137:349
57. Kostromin SG, Sinitzyn VV, Talroze RV, Shibaev VP (1982) Macromol Chem Rapid Commun 3:809
58. Shibaev VP, Plate NA (1985) Pure Appl Chem 57:1589
59. Shibaev VP, Kostromin SG, Plate NA (1982) Eur Polym J 18:651
60. Hsu CS, Percec V (1987) Polym Bull 18:91
61. Cho I, Chang K (1997) Macromol Rapid Commun 18:45
62. Komiya Z, Schrock RR (1993) Macromolecules 26:1387
63. Viertler K, Wewerka A, Noirez L, Stelzer F (2002) In: Khosravi E, Szymanska-Buzar T (eds) Ring opening metathesis polymerisation and related chemistry (Nato Science Series, II: Mathematics, Physics and Chemistry), vol 56. Kluwer Academic, Dordrecht, p 143
64. Maughon BR, Weck M, Mohr B, Grubbs RH (1997) Macromolecules 30:257

65. Riegler S (2003) PhD Thesis. Graz University of Technology, Graz
66. Horvath IT (1998) Acc Chem Res 31:641
67. Wewerka K, Wewerka A, Stelzer F, Gallot B, Andruzzi L, Galli G (2003) Macromol Rapid Commun 24:906
68. Koltzenburg S, Ungerank M, Stelzer F, Nuyken O (1999) Macromol Chem Phys 200:814
69. Pugh C, Schrock RR (1993) Polym Prep 34:180
70. Pugh C, Arehart S, Liu H, Narayanan R (1994) Pure Appl Chem A 31:1591
71. Russell TP, Rabolt JF, Twieg RJ, Siemens RL, Farmer BL (1986) Macromolecules 19:1135
72. Pugh C, Dharia J, Arehart SV (1997) Macromolecules 30:4520
73. Kim G-H, Pugh C, Cheng SZD (2000) Macromolecules 33:8983
74. Kim G-H, Jin S, Pugh C, Cheng SZD (2001) J Polym Sci Pol Phys 39:3029
75. Pugh C, Shao J, Ge JJ; Cheng SZD (1998) Macromolecules 31:1779
76. Arehart SV, Pugh C (1997) J Am Chem Soc 119:3027
77. Small AC, Hunt DK, Pugh C (1999) Polym Prepr 40:526
78. Small AC, Pugh C (2002) Macromolecules 35:2105
79. Pugh C, Bae J-Y, Dharia J, Ge JJ, Cheng SZD (1998) Macromolecules 31:5188
80. Pugh C, Thompson MJ, Mullins RJ, Hwang JH (1999) Polym Prepr 40:536
81. Li M-II, Keller P, Albouy P-A (2003) Macromolecules 36:2284
82. Weck M, Mohr B, Maughon BR, Grubbs RH (1997) Macromolecules 30:6430
83. Percec V, Chu P, Asandei AD (1999) Polym Mater Sci Eng 80:223
84. Percec V, Holerca MN (2000) Biomacromolecules 1:6
85. Percec V, Schlueter D (1997) Macromolecules 30:5783
86. Percec V, Schlueter D, Ronda JC, Johansson G, Ungar G, Zhou JP (1996) Macromolecules 29:1464
87. Walba DM, Keller P, Shao R, Clark NA, Hillmyer M, Grubbs RH (1996) J Am Chem Soc 118:2740
88. Joo S-H, Yun Y-K, Jin J-I, Kim D-C, Zin W-C (2000) Macromolecules 33:6704
89. Joo S-H, Jin J-I (2004) J Polym Sci Pol Chem 42:1335
90. Demel S, Slugovc C, Stelzer F, Fodor-Csorba K, Galli G (2003) Macromol Rapid Commun 24:636

Received: June 2004

Regioselective Polymerization of 1-Alkynes and Stereoselective Cyclopolymerization of α,ω-Heptadiynes

Michael R. Buchmeiser (✉)

Arbeitskreis Makromolekulare Chemie, Institut für Analytische Chemie und Radiochemie, Leopold-Franzens-Universität Innsbruck, Innrain 52a, 6020 Innsbruck, Austria
michael.r.buchmeiser@uibk.ac.at

1	Introduction	90
2	Polymerization of Ferrocene- and Ruthenocene-Substituted 1-Alkynes	92
3	Alternating 1-Alkyne-Ring-Opening Metathesis-Polymerization	95
4	Spaced Metallocenylalkynes	96
4.1	(Ferrocenylethynyl)-4′-ethynyltolan: A Tailor-Made 1-Alkyne for Mechanistic Investigations	96
4.2	Tailor-Made Spaced Alkynes for Main Chain Tuning	97
5	Cyclopolymerization of 1,6-Diynes	100
5.1	Polymer Structure	101
5.2	Stereoselective Cyclopolymerization	103
5.2.1	Diethyl Dipropargylmalonate (DEDPM)	103
5.2.2	Polymerization of Chiral Monomers	106
5.3	Livingness	108
5.4	Initiators Based on M-n-Bu$_4$Sn-EtOH-Quinuclidine (M=MoCl$_5$, MoOCl$_4$)	108
5.5	Ruthenium-Based Cyclopolymerization Systems	109
5.6	Supported Ruthenium-Based Cyclopolymerization Systems	112
5.7	Physical Properties of Poly(DEDPM)$_n$	114
5.7.1	Poly(DEDPM)$_n$ Based on Cyclopent-1-enylene-1-vinylenes	114
5.7.2	Poly(DEDPM)$_n$ Based on Cylohex-1-ene-3-methylidenes	115
6	Conclusions	115
	References	116

Abstract Metathesis-based polymerizations of 1-alkynes and cyclopolymerizations of 1,6-heptadiynes using late transition metal catalysts are reviewed. Results obtained with both binary, ternary, and quaternary catalytic systems and well-defined molybdenum- and ruthenium-based catalysts are presented. Special consideration is given to advancements in catalyst design and mechanistic understanding that have been made in this area over the last few years; advancements that have facilitated tailor-made syntheses of poly(ene)s. In addition, the first supported ruthenium-based cyclopolymerization-active systems are summarized. Finally, selected structure-dependent properties will be outlined where applicable.

Keywords Metathesis · Polymerizations · Metallocenes · Homogeneous catalysis · Heterogeneous catalysis

List of Abbreviations
ADMET polymerization	acyclic diene metathesis polymerization
CT	charge transfer
HOMO	highest occupied molecular orbital
LUMO	lowest unoccupied molecular orbital
PDI	polydispersity
TCDTF6	7,8-bis(trifluoromethyl)tricyclo [4.2.2.0^{2,5}]deca-3,7,9-triene
THF	tetrahydrofurane
N_{eff}	effective conjugation length
ROMP	ring-opening metathesis polymerization
λ_{max}UV	absorption maximum
o-TMSPA	o-trimethylsilylphenylacetylene
IS	isomer shift (Moessbauer)
DEDPM	diethyl dipropargylmalonate
SDS	sodium dodecylsulfate

1
Introduction

Fully conjugated polymers based on poly(ene)s and related structures have many potential applications in the fields of organic (semi-) conductors, optoelectronics, and photonics [1–6]. Unfortunately, poly(acetylene), the simplest representative of this class of compounds, suffers from insolubility, lack of processability, and insufficient oxygen stability. Substituted poly(acetylene)s overcome these problems, and are accessible via three different routes. The first, ADMET (acyclic diene metathesis) polymerization, is a metathesis-based condensation reaction of α,ω-dienes [7, 8] and is widely used in syntheses of poly(p-phenylenevinylene)s (PPVs) and their derivatives [9–15]. This polymerization technique is characterized by a comparable easy set-up, and many catalysts are suitable for metathesis polymerization in principle [14]. Nevertheless, the current preference is for well-defined catalytic systems, in order to obtain well-defined polymers and to avoid comparably low molecular weights [16]. A detailed description of this polymerization method is given in the chapter by Baughman and Wagener. The second approach to poly(ene)s, ROMP (ring-opening metathesis polymerization), is restricted to specifically designed poly(ene) precursors such as substituted and unsubstituted cyclooctatetraenes or paracyclophenes [13–23]. The third, 1-alkyne polymerization, allows a wide range of structural variations in principle, once a suitable initiator monomer system has been identified (Scheme 1) [17–24].

Although classic ternary systems based on molybdenum, tungsten and rhodium have been widely used for these purposes [25–52], *terminal* alkynes are still best polymerized by well-defined *Schrock*-type metathesis initiators in

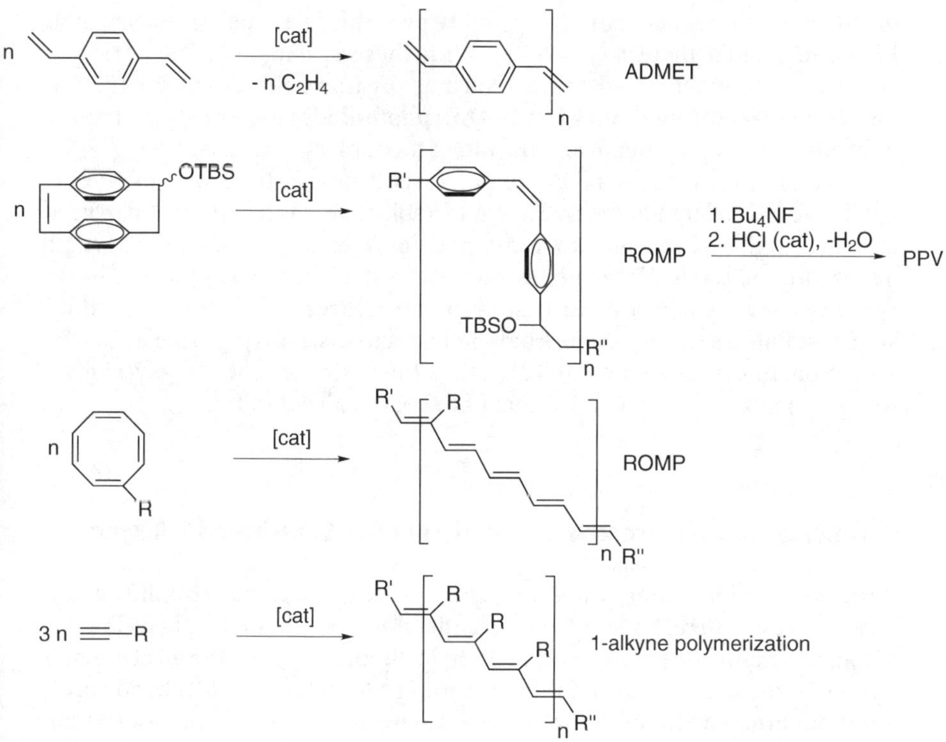

Scheme 1 Metathesis-based polymerization techniques for the synthesis of conjugated materials

order to yield conjugated poly(ene)s. Where a suitable initiator is found, these polymerizations can be carried out in a living manner [21, 23, 53–55]. In the case of 1,2-disubstituted alkynes, polymerizations are restricted to the use of binary and ternary systems, respectively [56–68]. Generally speaking, the degree of conjugation in conjugated materials strongly depends on the steric nature of the substituents in the starting alkyne. The desired coplanarity of the double bonds, a prerequisite for most applications, is best described by the effective conjugation length (N_{eff}). This is a measure of the number of coplanar double bonds in a conjugated system resulting from overlap of the p_z-orbitals. Highly conjugated systems show narrow bandgaps between HOMO and LUMO, which result in low-energy charge-transfer (CT) bands. As a consequence, these materials exhibit strongly bathochromic shifts in absorption. In 2000, recent achievements made in this area of research were acknowledged by awarding the Nobel Prize in Chemistry to H. Shirakawa, A.G. MacDiarmid and A.J. Heeger for their discovery and development of electrically-conducting polymers [3–5]. Despite their outstanding and promising role in the development of conjugated materials, poly(ene)s based on poly(acetylene)s were not able to fulfill these high expectations. Consequently, conjugated materials are currently generally based on poly(thiophene)s, poly(pyrrole)s, poly(thiazole)s, poly(p-phenylene)s, PPV and related materials [69–71]. With these systems, a large variety of conjugated polymer-based devices, such as organic field transistors [72, 73], diodes, light-emitting diodes [70, 74–76], photodiodes, polymer grid triodes, light emitting electrochemical cells, and optocouplers are available [71, 77, 78].

A major reason for the failure of poly(acetylene)s in the above-mentioned applications is related to their inherent instability versus moisture and oxygen, and their high susceptibility to decomposition/rearrangement in the partially oxidized/doped state. Nevertheless, poly(ene)s stabilized by appropriate ligand systems and/or incorporated into cyclic structures are believed to exhibit similar stabilities to poly(thiophene)s, poly(pyrrole)s, poly(p-phenylene)s, PPV, and so on. In the following, we will outline the basic concepts of poly(ene)s as well as reviewing the structures that have been realized so far.

2
Polymerization of Ferrocene- and Ruthenocene-Substituted 1-Alkynes

Besides other intriguing properties, such as inherent planar chirality, metallocenes are of interest due to their considerable Lewis basicity. Their direct or conjugative attachment to a polymer chain should enhance the electron density along the main chain and therefore lower the HOMO–LUMO band gap. In addition, organometallic compounds and metallocene-based monomers and polymers represent interesting potential nonlinear optical materials, useful for frequency doubling, modulation, and switching, for three reasons:

i) The compounds show metal-ligand (M-L) and L-M charge transfer (CT) bands in the UV-vis region. These transitions are very often associated with large second order nonlinearities.
ii) Intense colors (and therefore high transition dipole moments) are responsible for high second order nonlinearities.
iii) Since metallocenes are stable in different oxidation states, and β is expected to increase as $N^{3.05}$ (N is the number of carbons in the chain) for a charged species, cationic and anionic monomers and polymers would appear to be promising compounds. In particular, γ shows a high dependency on N ($\sim N^q$ with $5.0 < q < 5.25$) in charged systems.

Taking these features into consideration, polymerizations of the simplest metallocenyl alkynes (ferrocenyl- and ruthenocenylacetylene) were chosen to start with. Many attempts to polymerize ferrocenylacetylene (or more attractively ethynylferrocene) in a controlled way have been reported [79, 80]. Nevertheless, none of the systems investigated allowed a high yield synthesis of poly(ethynylferrocene) with satisfying physical properties (especially polydispersity (PDI), solubility, and molecular weight). In all cases, low molecular weight oligomers with high PDIs (5–10) were obtained as mostly insoluble, brown, probably partially oxidized, barely characterized materials. No relevant information about the general polymer structure or possible crosslinkage was accessible at all. In view of the encouraging results from 1-alkyne polymerization obtained with well-defined Mo-based initiators [17, 18, 20, 21, 81], usually referred to as "Schrock initiators", an extension of this chemistry to the preparation of metallocenyl-substituted poly(ene)s appeared promising.

Ethynylferrocene and ethynylruthenocene are, due to the Lewis base character of the metallocene moiety, highly reactive terminal acetylenes. In contrast to their phenyl analogues, they can (in principle) be polymerized using a wide variety of classic Schrock initiators. Nevertheless, in order to obtain a well-defined polymerization system that permits access to tailor-made polymers, one needs to bear in mind the two possible reaction pathways for 1-alkyne polymerization (Scheme 2).

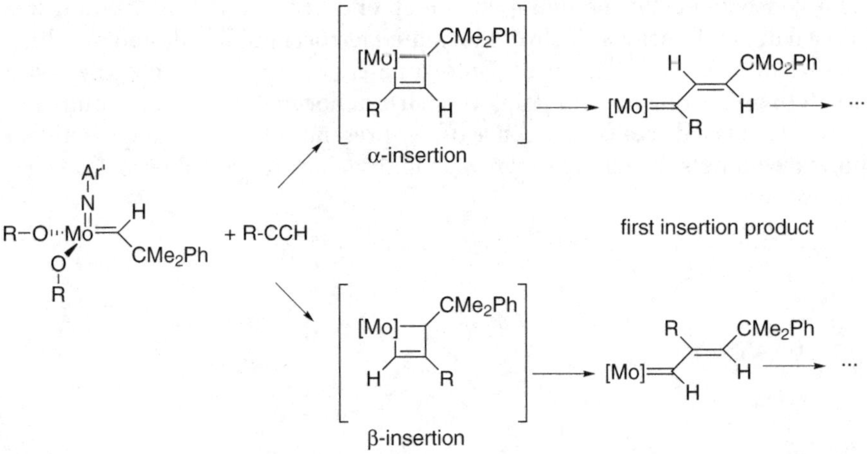

Scheme 2 Two different modes of insertion into 1-alkyne polymerization

The monomer can insert into the metal–carbene double bond via α- or β-insertion. In principle, both insertion modes result in the formation of a conjugated poly(ene). Nevertheless, in particular with highly reactive monomers, β-insertion often results in the formation of ill-defined polymers because of the subtle differences in the reactivities of the initiator and the first insertion products. If the rate constant for polymerization (k_p) is much larger than the rate constant for initiation (k_i) in these systems, only a low percentage of the initiator forms a propagating polymer chain, which directly results in a loss of control over the molecular weight. One observes the precipitation of virtually insoluble polymer from the reaction mixture within a few seconds after initiation. At this point it should be mentioned that the ratios of k_i/k_p are conveniently determined via ^{1}H-NMR [82] when the polymerization system fulfills the criteria of a truly "living" system [83–85]. In addition, one can easily distinguish between these two mechanisms by applying ^{1}H-NMR. When the polymerization proceeds via α-addition, disappearance of the starting alkylidene and formation of (usually *trans*) coupled signals is observed (Scheme 2). On the other hand, new alkylidene signals as well as isolated, non-coupled signals in the olefinic region appear when polymerization proceeds via β-addition. Based on the concept of "small alkoxides" developed by Schrock [21], where initiators that contain small alkoxide ligands favor α-addition due to the absence of steric interaction between the monomer and the alkoxides, $Mo(N-2,6-Me_2-C_6H_3)$ $(CHCMe_2Ph)(OCMe(CF_3)_2)_2$ turned out to be the best initiator among many other investigated systems. It allowed the synthesis of homo- as well as co-polymers in high yields and with high selectivity in terms of insertion mode and consequently head-to-tail connectivity [53]. The polymerization systems turned out to be truly living, facilitating syntheses of the corresponding poly(ene)s with required molecular weights and with PDIs typically <1.3. As a consequence of the living character, end-group functionalization was carried out with various aldehydes including ferrocene aldehyde and pyridine-4-aldehyde (see also Scheme 3). Unfortunately, values for N_{eff} obtained with poly(ethynylferrocene) and poly(ethynylruthenocene) in *THF* were quite low (<10). This is a direct consequence of the presence of bulky groups in linear poly(acetylene)s, which often prevent high values for N_{eff} due to 1,3-steric interactions.

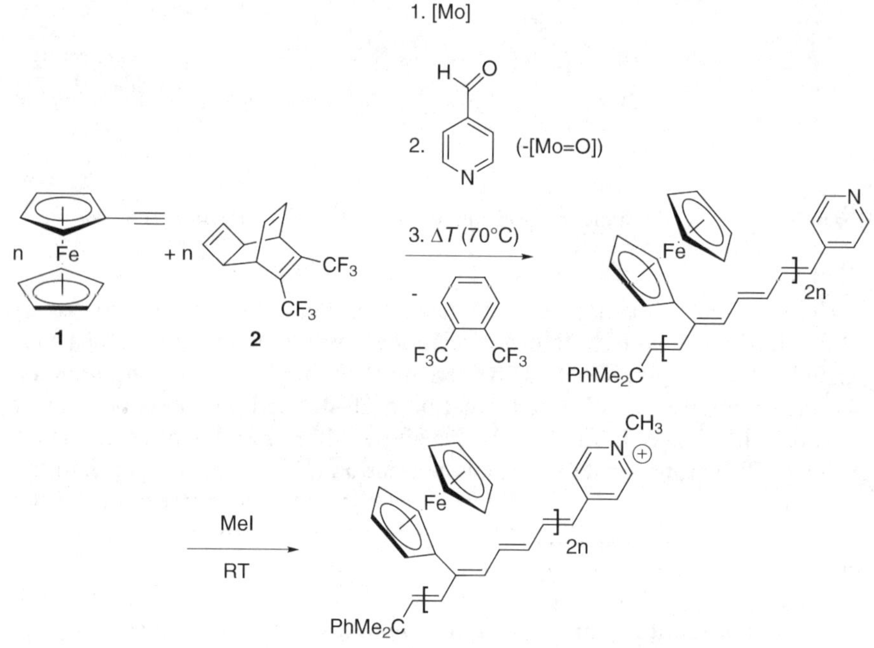

Scheme 3 Synthesis of ferrocene-substituted poly(ene) by alternating copolymerization

3
Alternating 1-Alkyne-Ring-Opening Metathesis-Polymerization

As an alternative, copolymerization of alkynes bearing bulky substituents with TCDTF6 (7,8-bis(trifluoromethyl)tricyclo [$4.2.2.0^{2,5}$]deca-3,7,9-triene) was carried out. In the course of this copolymerization, usually referred to as the "Durham Route" [86–89], the "Feast-monomer" was introduced into the polymer main chain and subsequently converted into three unsubstituted, conjugated double bonds via a thermally-induced retro-Diels Alder reaction (Scheme 3) [53].

Sufficient separation of one ferrocenyl substituent from the other is achieved for perfectly alternating copolymers. Ferrocene-substituted alkynes prepared by this route exhibit a remarkable stability due to steric protection by the bulky ferrocene substituents, despite the large number of unsubstituted double bonds. This translates directly into a significant bathochromic shift in absorption, which can even be enhanced by endgroup functionalization (quaternization). This quaternization which is carried out by simply treating with methyl iodide, results in the formation of highly conjugated, charged poly(ene)s (λ_{max}=595 nm, N_{eff}=54, $CHCl_3$), which exhibit a strong solvatochromic behavior. The preparation of such perfectly alternating copolymers using ROMP and 1-alkyne polymerization strongly depends on the ratios $k_{i(Ai)}/k_{i(Bi)}$ as well as $k_{p(Ai)}/k_{p(Bi)}$, where $k_{i(Ai)}$, $k_{i(Bi)}$, $k_{p(Ai)}$, and $k_{p(Bi)}$, and so on, represent the rates of insertion and the rates of propagation of the two monomers A (ethynylferrocene) and B (TCDTF6) at the corresponding insertion step i (i=1,2,3,...DP=degree of polymerization) with the active carbene, respectively. Ideally, $k_{i(Ai)} \gg k_{i(Bi)}$ if the last monomer that was inserted was B and vice versa. Unfortunately, these rate constants are difficult to control, significantly limiting the applicability of this concept.

As a direct consequence, the synthesis of highly conjugated poly(ene)s via the preparation of homopolymers from a single functional monomer still seems highly preferable. In order to do this, the substituted alkyne must fulfill at least two requirements in order to be suitable for the synthesis of poly(ene)s with high values of N_{eff}. On the one hand, the substituent should possess electron-donating character in order to activate the carbon-carbon triple bond and make it more reactive towards a transition metal alkylidene, which usually speeds up initiation. In addition, Lewis base character provides an enhanced electron-density in the resulting polymer backbone. On the other hand, the substituent should ideally be designed in such way that the conjugated double bonds in the resulting poly(ene) are as coplanar as possible. In principle this may be achieved by reducing the steric interaction between the substituents (by introducing spacers).

4
Spaced Metallocenylalkynes

4.1
(Ferrocenylethynyl)-4′-ethynyltolan: A Tailor-Made 1-Alkyne for Mechanistic Investigations

As already briefly mentioned, Schrock et al. proposed that the use of "small", electron-withdrawing alkoxides favors α-addition, whereas larger alkoxides should give rise to β-addition due to steric hindrance. This concept proved to be useful, particularly for the polymerization of o-substituted phenylacetylenes, such as o-trimethylsilylphenylacetylene (o-TMSPA) and related compounds [21, 23], although other terminal acetylenes (*like* the above-mentioned compounds ethynylferrocene and ethynylruthenocene) behaved in a different way (adding via α-insertion despite the use of large alkoxides in $Mo(N\text{-}2,6\text{-}Me_2C_6H_3)(CHCMe_2Ph)(OCMe(CF_3)_2)_2$, and vice versa with metal carbenes containing small alkoxides such as $Mo(N\text{-}2,6\text{-}Me_2C_6H_3)(CHCMe_2Ph)(OC_6F_5)_2 \cdot$quinuclidine) [53]. Evidently and not surprisingly, not only the initiator but also the monomer plays an important role in terms of sterics and electronics. Therefore, acetylenes containing electron-withdrawing groups do not readily react, even with one of the most reactive molybdenum-based initiators, $Mo(N\text{-}2,6\text{-}iso\text{-}Pr_2C_6H_3)(CHCMe_2Ph)(OCMe(CF_3)_2)_2$. Interestingly enough, nor do other either electron-rich, yet sterically-hindered acetylenes, such as *tert*-butylacetylene. As the ferrocene moiety obviously possesses some special properties with regard to the reactivity of the monomer itself, the polymerization of a tailor-made monomer, 4-(ferrocenylethynyl)-4′-ethynyltolan, a diethynyltolan-spaced homologue of ethynylferrocene, appeared promising for two reasons (Fig. 1).

First, this compound was sterically demanding and should therefore, if there is any reactivity at all, undergo β-addition. Second, it contained the ferrocene moiety, yet the reactivity of the terminal alkyne group was reduced because of the ethynyltolan spacer. As expected, the monomer was significantly less reac-

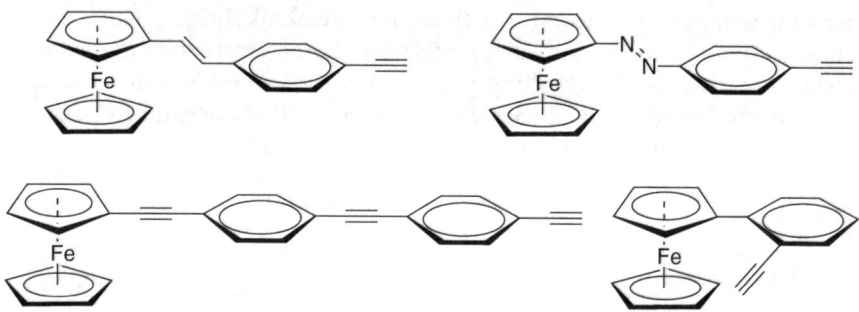

Fig. 1 Structures of ferrocene-substituted alkynes

tive than ethynylferrocene, as the additional electron density was distributed over a larger π-system. As anticipated, this compound polymerized via β-addition using Mo(N-2,6-iso-Pr$_2$C$_6$H$_3$)(CHCMe$_2$Ph)(OCMe(CF$_3$)$_2$)$_2$ as the initiator [90]. Evidence for this type of insertion was provided by ^{1}H NMR spectroscopy. Upon treatment of the initiator with 1 equiv of the monomer, a sharp singlet for H$_\gamma$ appeared at 5.93 ppm. No coupling was observed. The signal for the terminal alkyne at 3.17 ppm disappeared and a new alkylidene resonance for H$_\alpha$ at 12.37 ppm appeared. Surprisingly, no signals for a second or third insertion product were observed, indicating a ratio of $k_p/k_i \leq 1$. Beside the signals for H$_\alpha$ and H$_\gamma$, only one characteristic signal group for the ferrocene moiety of the first insertion product was observed. These NMR data were in accord with the living character of the polymerization, which was confirmed for the first time for a 1-alkyne polymerization that proceeded via β-addition, indicating the importance of kinetic parameters (k_p/k_i) as well as insertion issues for controlling the molecular weight. Consequently and in contrast to other polymerizations proceeding via β-addition, the resulting materials containing up to 50 metallocene groups possessed excellent solubility and could be prepared with control over molecular weight and PDI (generally <1.3). For the first time there was access to the controlled synthesis of poly(ene)s containing large side groups. The remaining question was whether this mode of insertion and polymerization behavior applied to other spaced 1-alkynes too.

4.2
Tailor-Made Spaced Alkynes for Main Chain Tuning

To answer this question, the conjugated ferrocenylacetylenes 2-(4-ethynylphen-1-yl)-vin-1-ylferrocene, 1-ferrocenyl-2-(4-ethynylphenyl)diazene, and 2-ethynylphenylferrocene were synthesized (Fig. 1) [54].

The length and structure of these spacers had significant influence on the mode of insertion of the corresponding alkyne into the Schrock-carbene as well as on the structure of the resulting poly(ene). Both 2-(4-ethynylphen-1-yl)-vin-1-ylferrocene and 1-ferrocenyl-2-(4-ethynylphenyl)diazene underwent selective β-addition with Mo(N-2,6-Me$_2$-C$_6$H$_3$)(CHCMe$_2$Ph)(OCMe(CF$_3$)$_2$)$_2$ and Mo(N-2,6-iso-Pr$_2$-C$_6$H$_3$)(CHCMe$_2$Ph)(OCMe(CF$_3$)$_2$)$_2$, respectively. This was not surprising, since their molecular structures are quite similar to that of 4-(ferrocenylethynyl)-4'-ethynyltolan (Fig. 1). Interestingly, 2-ethynylphenylferrocene turned out to be a sensitive probe for the steric (and electronic) situation of an initiator. Therefore, polymerization proceeded via α-addition with Mo(N-2,6-Me$_2$-C$_6$H$_3$)(CHCMe$_2$Ph)(OCMe(CF$_3$)$_2$)$_2$, while β-addition was observed with Mo(N-2,6-iso-Pr$_2$-C$_6$H$_3$)(CHCMe$_2$Ph)(OCMe(CF$_3$)$_2$)$_2$. Since both initiators were based on identical alkoxides, an additional factor had to be responsible for this observation. The transition state of both insertion modes provided some useful insight (Fig. 2).

We can deduce from this that the 2-ferrocenylphenyl group gains maximum distance from the initiator end group (-CMe$_2$Ph) at the expense of steric inter-

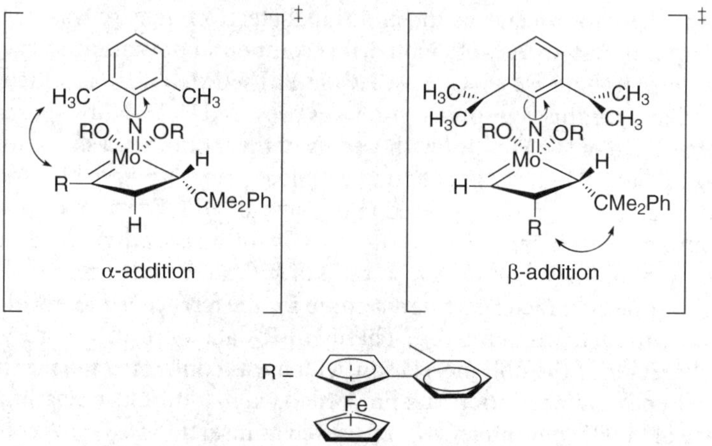

Fig. 2 Influence of aryl substituents on the mode of insertion in the polymerization of sterically-demanding 1-alkynes

actions with the comparably small methyl groups (chain end controlled, α-addition). In contrast, this is no longer possible in the case of the "large" isopropyl groups; the reaction is initiator controlled and proceeds via β-addition. This only applies to o-substituted phenylalkynes with no steric interactions of the ferrocene substitutent with the hexafluoro-*tert*-butoxide group.

k_i was at least comparable to k_p for all three monomers, leading to truly living polymerization with good control over molecular weight. Polymers prepared from the linearly spaced monomers 2-(4-ethynylphen-1-yl)-vin-1-ylferrocene and 1-ferrocenyl-2-(4-ethynylphenyl)diazene possessed low effective conjugation (N_{eff}<10, THF), while poly(2-ethynylphenylferrocene) showed a significant bathochromic shift in absorption with a λ_{max} of 515 nm, corresponding to an N_{eff} of 17 (THF). The rather high value for N_{eff} for 2-trimethylsilylphenylacetylene [21, 23] compared to other metallocene-substituted poly(ene)s was suggested to be due to π-stacking [91, 92] of the adjacent phenyl rings, but this explanation requires final proof.

In order to enhance the effective conjugation lengths, a series of octamethyl-derivatized ferrocene-based 1-alkynes were synthesized (Fig. 3).

The monomers [2- [4-(ethynyl)-phenyl]-ethenyl]-1′,2,2′,3,3′,4,4′,5-octamethylferrocene, [2- [(5-ethynyl)-thien-2-yl]-ethenyl]ferrocene, [2- [(5-ethynyl)-thien-2-yl]-ethenyl]-1′,2,2′,3,3′,4,4′,5-octamethylferrocene, (*E*)-2-(ethynylferrocenyl)-ethenyl]-1′,2,2′,3,3′,4,4′,5-octamethylferrocene and [2-(2-ethynylphenyl)-ethenyl]-1′,2,2′,3,3′,4,4′,5-octamethylferrocene all polymerized via β-addition with Mo(*N*-2,6-Me$_2$-C$_6$H$_3$)(CHCMe$_2$Ph)(OCMe(CF$_3$)$_2$)$_2$ as the initiator. Again only the o-substituted analogue to 2-ethynylphenylferrocene, [2-(2-ethynylphenyl)-ethenyl]-1′,2,2′,3,3′,4,4′,5-octamethylferrocene, facilitated the synthesis of conjugated poly(ene)s with N_{eff} values of 20 (THF). In order to shed some light on whether the proposed Lewis base character of the pendent

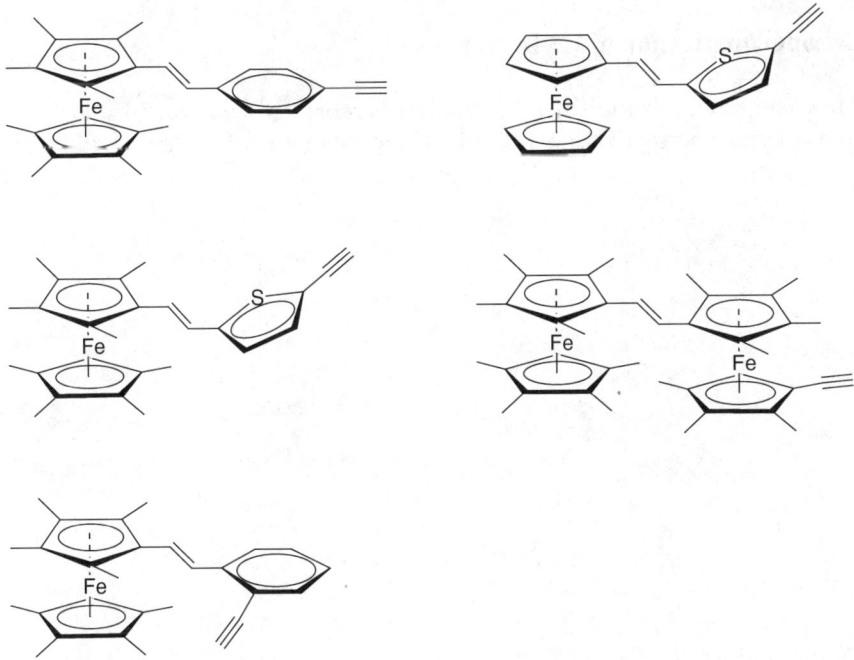

Fig. 3 Structure of octamethylferrocene-substituted 1-alkynes and [2-[(5-ethynyl)-thien-2-yl]-ethenyl]-ferrocene

or spaced ferrocenes enhances the electron density along the poly(ene) backbone, ^{57}Fe-Mössbauer experiments were carried out. Aside from the quadrupole splitting, the isomer shift (IS) is a good indicator of electron density around the metal core. If electron density is transferred to the backbone, a positive deviation from the standard, ferrocene and ethynylferrocene, respectively, is observed. *In fact, investigations revealed a significant positive deviation from the standard ISs, indicative of electron transfer to the backbone* [93]. Therefore, data obtained by ^{57}Fe-Mössbauer spectroscopy reflect a small decrease in s-electron density within the ferrocene moiety and its transfer to the conjugated backbone, as demonstrated by the reduced ISs of the monomers ethynylferrocene, 2-ethynylphenylferrocene, [2- [ethynylphen-2-yl]ethen-1-yl]-1′,2,2′,3,3′,4,4′,5-octamethylferrocene compared to the corresponding polymers. In addition, the temperature dependence of the recoil-free fraction and the calculated lattice temperatures indicated that the polymers were much "softer" (within the definitions of Mössbauer spectroscopy) than the corresponding monomers.

5
Cyclopolymerization of 1,6-Diynes

The cyclopolymerization of 1,6-heptadiynes represents a powerful alternative to 1-alkyne polymerization [81, 94]. Cyclopolymerizations may be accom-

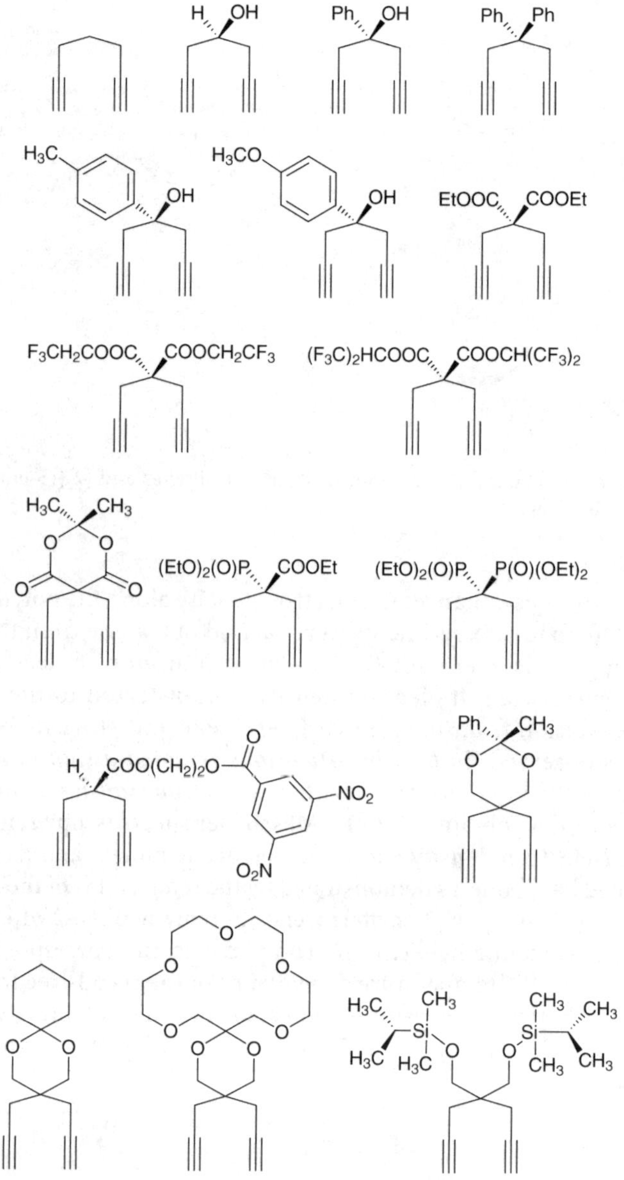

Fig. 4 1,6-Heptadiyne derivatives used in cyclopolymerization

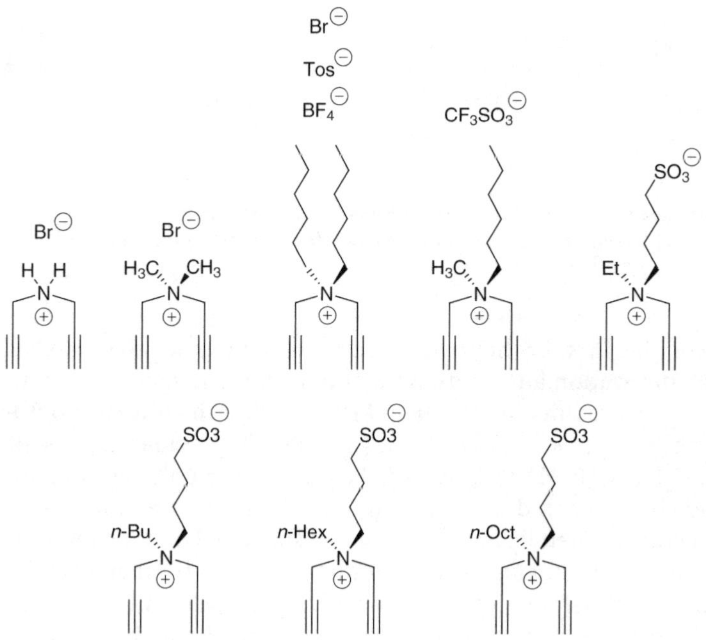

Fig. 5 Charged 1,6-heptadiyne derivatives used in cyclopolymerization

plished using Ziegler-Natta-catalysts [95–100], Pd-catalysts [101], anionic polymerization [102], binary and ternary Mo- or W-based catalysts [103] and Ni-based catalysts [104]. $MoCl_5$-derived systems were more reactive than the corresponding WCl_6-based in Terms of with Group VI-based initiators [105]. In the case of unsubstituted diynes such as 1,2-diethynylbenzene or 1,6-heptadiyne, most of these catalytic systems lead to mostly brown, insoluble, ill-defined polymers with variable repetitive units and broad molecular weight distributions. Introduction of a ligand in the 4-position generally improves the solubility of the resulting polymers. Therefore, substituted heptadiynes possess good solubilities in common organic solvents (C_6H_6, toluene, CH_2Cl_2, $CHCl_3$), good long-term stabilities towards oxidation, and high effective conjugation lengths [24, 81, 103, 106]. Cyclopolymerizations of a large number of uncharged [24, 81, 94, 106–118] and ionic monomers with substituents at the 4-, 2- and 5-position have been investigated [119–124]. In most of these investigations, binary and ternary catalysts have been employed. These monomers are reviewed in Figs. 4 and 5.

5.1
Polymer Structure

In terms of polymer structure, cyclopolymerizations of 1,6-heptadiynes usually yield poly(ene)s that contain a mixture of five- and six-membered rings (Fig. 6).

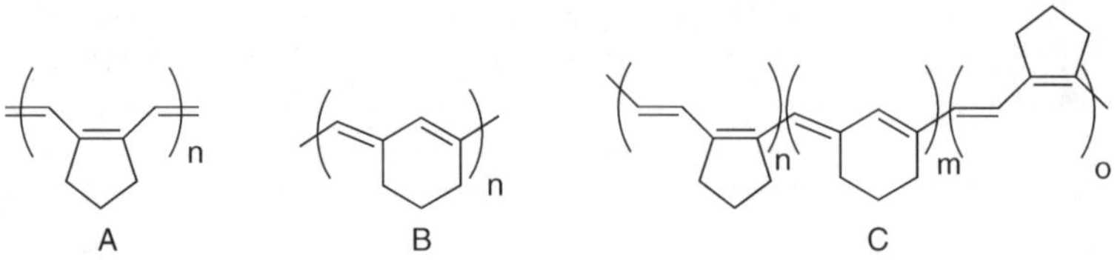

Fig. 6 Possible ring structures of poly(1,6-heptadiynes) prepared via cyclopolymerization. Poly(cyclopent-1-enylene-1-vinylene)s (**A**); poly(cyclohex-1-ene-3-methylidene)s (**B**); and mixed structures (**C**)

One of the first 1,6-heptadiyne-derived monomers that was subject to cyclopolymerization and whose structure was unambiguously identified was diethyl dipropargylmalonate (DEDPM) [81, 125], which turned out to be *the* working horse in this research area [24, 81, 106]. Its polymerization with Mo(N-2,6-*i*Pr$_2$-C$_6$H$_3$)(CHCMe$_2$Ph)(OCMe(CF$_3$)$_2$)$_2$ yielded a polymer containing both five- and six-membered ring structures. It is important to note that these two ring structures generally occur within one single polymer chain, and not in a mixture of polymer chains consisting either of five- or six-membered rings. The structure of this polymer was determined by ^{13}C-NMR, since the carbonyl carbon and the quaternary carbon are especially sensitive to the ring size. Carbon resonances at 172.0 ppm (carbonyl carbon) and 57–58 ppm (quaternary carbon) are characteristic of five-membered rings, while resonances at 170.8 and 54 ppm are characteristic of six-membered rings. Nevertheless, the *cis/trans* configura-

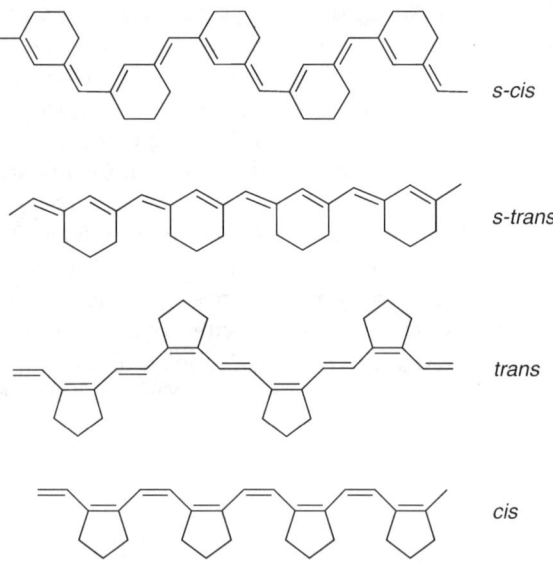

Fig. 7 Possible conformations of poly(cyclopent-1-enylene-1-vinylene)s and poly(cyclohex-1-ene-3-methylidene)s

tion of the exocyclic double bond or the *s-cis/s-trans* configuration of the backbone single bond in poly(cyclohex-1-ene-3-vinylene) and the configuration of the double bonds in poly(cyclopent-1-enylene-1-vinylene) cannot be determined, making it difficult to distinguish between the two possible idealized planar conformational isomers (Fig. 7).

The reasons for this are various elements of symmetry: mirror planes, axes of rotation, centers of inversion that can be found in poly(cyclopent-1-enylene-1-vinylene)s and which relate each repetitive unit to each other. In poly(cyclohex-1-ene-3-vinylene)s, these elements are missing, but the low values for $^4J_{HH}$ do not allow any statements on configuration.

5.2
Stereoselective Cyclopolymerization

5.2.1
Diethyl Dipropargylmalonate (DEDPM)

In order to understand the polymer structures that are obtained in the polymerization of 1,6-heptadiynes, one needs to consider all possible polymerization mechanisms. If 1,6-heptadiynes are subject to cyclopolymerization using well-defined Schrock catalysts, polymerization can proceed via two mechanisms. One is based on monomer insertion, where the first alkyne group adds to the molybdenum alkylidene forming a disubstituted alkylidene, which then reacts with the second terminal alkyne group to form poly(ene)s consisting of five-membered rings. Analogous to 1-alkyne polymerization, one refers to this type of insertion as α-insertion (Scheme 4).

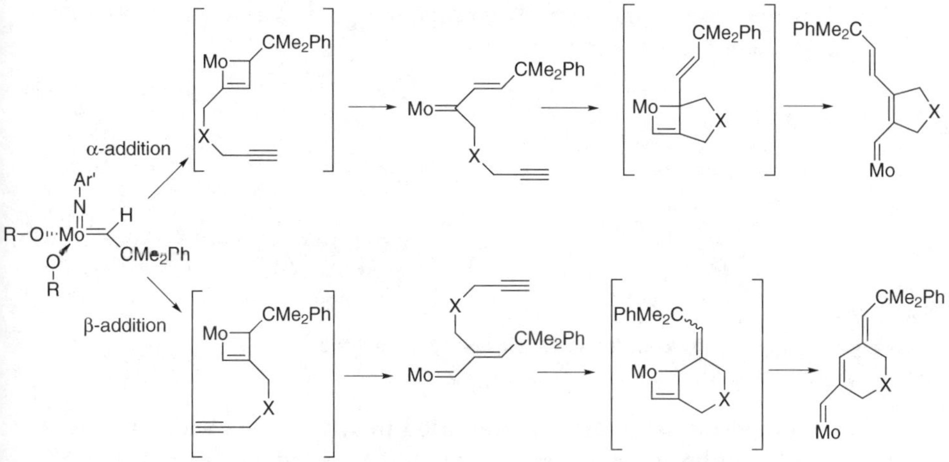

Scheme 4 Two possible reaction pathways and resulting ring structures in the cyclopolymerization of 1,6-heptadiynes

The monomer can also add to the initiator to form a monosubstituted alkylidene. This sequence, again referred to as β-addition, yields poly(ene)s based on six-membered rings (Scheme 2). Poly(DEDPM) *exclusively consisting of six-membered rings* was first prepared using the molybdenum imido alkylidene complex Mo(N-2-*tert*-Bu-C$_6$H$_4$)(CH-*tert*-Bu)(O$_2$CCPh$_3$)$_2$ [24, 106]. Here the bulky ligands (the carboxylates) forced the monomer to undergo selective β-addition by sterically preventing any other approach to the CNO-face [22]. These findings were in absolute accordance with the concept of "small alkoxides" developed by Schrock et al. for this type of initiator [21, 23]. Nevertheless, it should be emphasized that both Schrock et al. [126] and others [54] have already emphasized the important roles played by any substituents at the arylimido ligand, which can totally invert the mode of insertion for the same set of alkoxides [54]. The structure of poly(DEDPM), consisting solely of cyclohex-1-ene-3-vinylene units, was identified by ^{13}C-NMR as described above.

The polymerization of DEDPM in order to synthesize of poly(1,2-cyclopent-1-enylene-vinylene) required some significant tuning of the Schrock-type initiator. In order to obtain a polymer consisting virtually solely of cyclopent-1-ene-1-vinylene units, small alkoxides were required to allow α-addition. Unfortunately, the use of small alkoxides turned out to be a necessary, although not sufficient, prerequisite. In fact, some additional modifications in catalyst design had to be carried out. These modifications were related to some basic properties of molybdenum complexes of type Mo(NAr')(CHCMe$_2$Ph)(OR')$_2$. Generally speaking, these compounds exist in the form of two rotamers [22, 126]. One isomer, with the 2-phenylpropyl group pointing towards the imido ligand, is called the *syn*; the other, with this group pointing away from the imido group, is called the *anti* rotamer. These two rotamers interconvert with a rate that depends on both the electronic and steric effects around the Mo-center, allowing us to tune their reactivity and selectivity (Scheme 5) [22, 126–128]. Even more importantly, the *syn* and *anti* rotamers possess different reactivities, the *anti* rotamer being the more reactive.

Scheme 5 *syn-anti* interconversion of Schrock-type catalysts

These properties have already been used to synthesize a wide variety of stereoregular norborn-2-ene- and norbornadiene-based polymers [22, 54, 55, 126, 129]. Adding a base such as quinuclidine has a strong impact on these polymerizations. Although the base is not believed to coordinate to molybdenum during insertion [23], it strongly influences the reactivity of the entire system [18, 127]. The presence of a base, particularly at low temperatures, favors the formation of (coordinated) *anti* rotamers, since it stabilizes this isomer [126]. On the other hand, the presence of the base enhances the relative reactivity of the *syn* isomer. Fluoroalkoxide-based initiators such as Mo(N-2,6-Me$_2$-C$_6$H$_3$)(CHCMe$_2$Ph)(OCMe(CF$_3$)$_2$)$_2$ are not capable of forming poly(DEDPM) consisting solely of five-membered rings [81]. Neither addition of quinuclidine nor lower polymerization temperature significantly changed this situation. Since *syn-anti* interconversion is slow in these complexes, *the final geometry of a cyclopolymerization-derived polymer must at least be influenced by – if not governed by – the relative reaction rates of the syn and anti isomers and the rate of interconversion*. If this is true, initiators based on non-fluorinated alkoxides should be expected to facilitate the preparation of the target polymer, since they generally show fast *syn-anti* interconversion. As a matter of fact, Schrock initiators containing non-fluorinated alkoxides could be used for the living polymerization of DEDPM to produce poly(ene)s *based solely on five-membered rings* [130, 131]. Therefore, low-temperature-initiated polymerization of DEDPM using Mo(N-2,6-*iso*-Pr$_2$-C$_6$H$_3$)(CHCMe$_2$Ph)(OCH(CH$_3$)$_2$)$_2$ as the catalyst yielded virtually solely poly(cyclopent-1-enylene-1-vinylene). The same polymer was obtained using either Mo(N-2,6-Me$_2$-C$_6$H$_3$)(CHCMe$_2$Ph)(OC(CH$_3$)$_3$)$_2$·quinuclidine or Mo(N-2,6-*iso*-Pr$_2$-C$_6$H$_3$)(CHCMe$_2$Ph)(OC(CH$_3$)$_3$)$_2$·quinuclidine (Fig. 8). In these cases, the polymerizations were conducted at room temperature. Again, the polymer structure was assigned by ^{13}C-NMR, as described above.

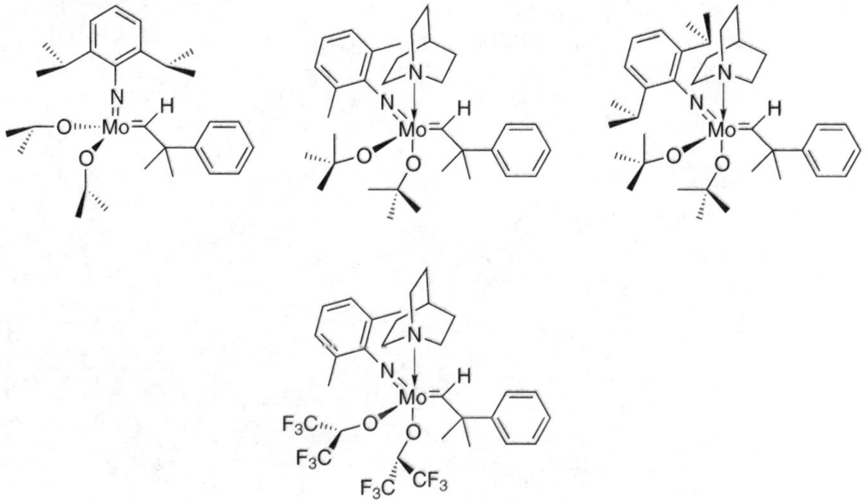

Fig. 8 Schrock-type catalysts successfully used in regioselective cyclopolymerization

5.2.2
Polymerization of Chiral Monomers

Chiral monomers (monomers containing at least one asymmetric carbon) allow us to retrieve extra information about the actual configuration of a poly(ene) and the relative orientations of the repetitive units (the tacticity). The chiral monomers di-(1S, 2R, 5S)-(+)-menthyl dipropargylmalonate and 4-(ethoxycarbonyl)-4-(1S, 2R, 5S)-(+)-menthoxycarbonyl-1,6-heptadiyne were cyclopolymerized to yield poly(ene)s consisting virtually solely of five-membered rings, using Mo(N-2,6-iso-Pr$_2$-C$_6$H$_3$)(CHCMe$_2$Ph)(OC(CH$_3$)$_3$)$_2$ and Mo(N-2,6-iso-Pr$_2$-C$_6$H$_3$)(CHCMe$_2$Ph)(OC(CH$_3$)$_3$)$_2$· quinuclidine, respectively. Generally, the use of such chiral monomers reduces the elements of symmetry one might detect during diade or triade interpretation (mirror planes, centers of inversion, centers of rotation) to one single element (center of rotation)

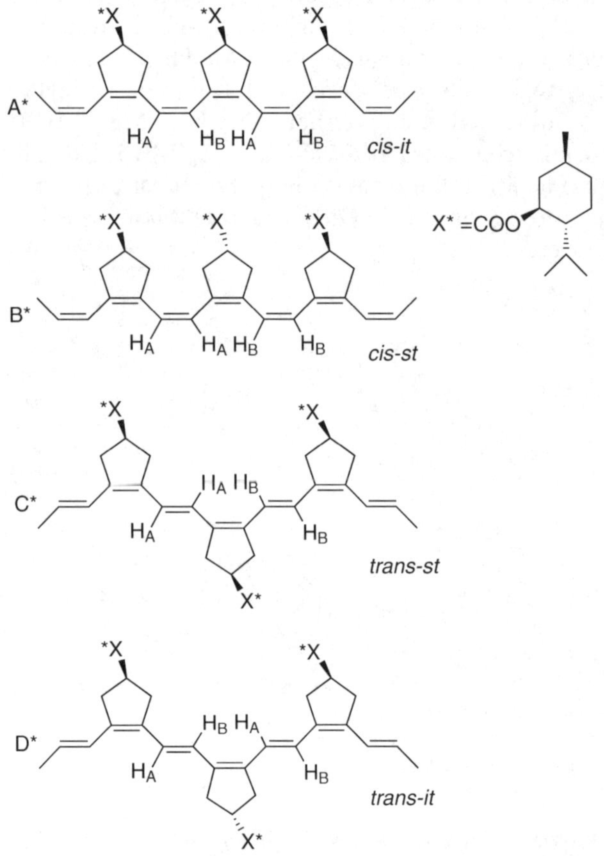

Fig. 9 Determination of double bonds in poly(cyclopent-1-enylene-1-vinylene)s containing chiral pendent groups

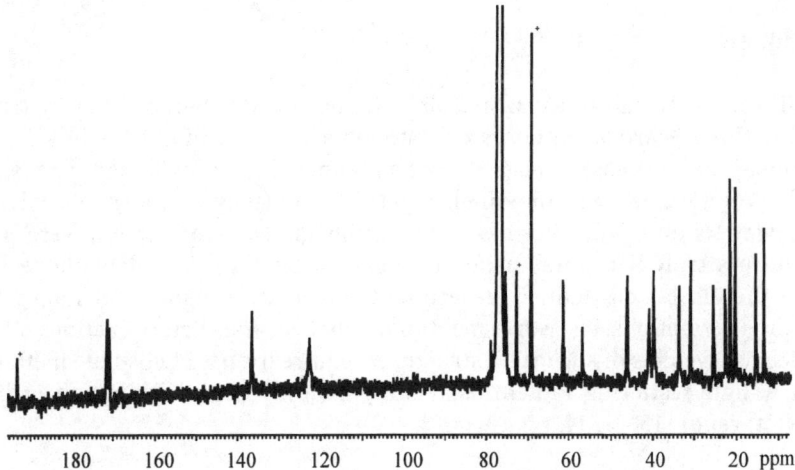

Fig. 10 ^{13}C-NMR spectrum of poly(4-(ethoxycarbonyl)-4-(1S, 2R, 5S)-(+)-menthoxycarbonyl-1,6-heptadiyne) based on >96% five-membered rings and containing an alternating *cis-trans st* structure. Signals indicated by an asterix (*) denote excess of ferrocene aldehyde used for initiator capping

(Fig. 9). Generally, no mirror planes or centers of inversion between any of the repetitive units can be found since such an element of symmetry would require a change in the absolute configuration of the asymmetric carbons (in this case in the pendent menthyl groups). Nevertheless, structures B* and C* (Fig. 9) possess an element of symmetry. In both cases, an axis of rotation can be found, one lying in the plane in the middle of the exocyclic double bond (B*) and one passing perpendicularly through the exocyclic double bond (C*). Therefore, the two protons and carbons, respectively, of the exocyclic double bond are chemically and magnetically equivalent and should therefore possess the same chemical shift. In addition, the two protons should not show any coupling.

The ^{13}C-NMR spectrum of poly(4-(ethoxycarbonyl)-4-(1S,2R,5S)-(+)-menthoxycarbonyl-1,6-heptadiyne) indicates a polymer containing almost solely five-membered rings (Fig. 10). In addition, only one single set for each type of carbon was observed, indicative of a highly tactic base. Keeping the symmetry restrictions described above in mind, either a *cis* or *trans-st* structure can be assigned.

A 500 MHz ^{1}H–^{1}H-correlated spectrum did not show any coupling between the olefinic protons, but a *trans*-structure could still be derived from analyzing the double bonds located next to the end groups. Therefore, an alternating *cis-trans* conformation was assigned to the polymer, which means that the polymer should possess a syndiotactic base. Nevertheless, since there could be some small, undetectable coupling, this assignment still remains speculative, although highly probable.

5.3
Livingness

While classical catalysts usually result in ill-controlled polymerization systems, well-defined Schrock initiators cyclopolymerize 1,6-heptadiynes in a living manner – in most cases a class VI living manner (according to Matyjaszewski [84]). Polymers based on one single repetitive unit (poly(cyclopent-1-enylene-1-vinylene)s and poly(cyclohex-1-ene-3-vinylene)s, respectively), were prepared in a truly living way using the Schrock catalysts described above and were therefore well defined in terms of molecular weights and molecular weight distributions. It is worth mentioning that accurate determinations of the molecular weights of such rigid structures require the use of absolute methods such as light scattering, rather than relative methods such as calibration against poly(styrene) [130, 131].

5.4
Initiators Based on M-n-Bu$_4$Sn-EtOH-Quinuclidine (M=MoCl$_5$, MoOCl$_4$)

An important aspect of the large-scale synthesis of any material is the cost. Despite their superiority, Schrock initiators are characterized by limited commercial availability and high sensitivity to oxygen and moisture. Therefore, classical ternary systems have been investigated but they have not yet been able to compete with Schrock systems in terms of the quality of the material produced (definition, purity, and so on). Nevertheless, there is still significant interest in the use of alternative systems to cyclopolymerize 1,6-heptadiynes. While standard ternary systems behaved much as expected, yielding ill-defined polymers without any control over molecular weight, the addition of quinuclidine turned out to be a milestone in this area of research [130, 131]. DEDPM was cyclopolymerized by MoCl$_5$-n-Bu$_4$Sn-EtOH-quinuclidine (1:1:5:1) and MoOCl$_4$-n-Bu$_4$Sn-EtOH-quinuclidine (1:1:2:1) to produce poly(DEDPM) *based exclusively on cyclopent-1-enylene-1-vinylene units*. The initiator efficiency of MoOCl$_4$-n-Bu$_4$Sn-EtOH-quinuclidine (1:1:2:1) was as high as 91%, the highest value ever reported for such systems, whereas the efficiency for the corresponding MoCl$_5$-based initiator was ≤67%. The absorption maximum λ_{max} for poly(DEDPM) was 587 nm, close to that found for poly(DEDPM) prepared by a Schrock initiator (592 nm, vide infra). A maximum effective conjugation length N_{eff} of 49 (THF) was calculated for this product. A plot of the number of monomers (N) added versus the molecular weight (M_n) as determined by light scattering showed a linear relationship for both initiators. Multistage polymerizations of DEDPM with both initiator systems indicated that the catalytic species were active for at least six hours in the presence of monomer, but they did not fulfill the criteria of a living polymerization (stability of the active chain end in the absence of monomer). Despite the fact that identical polymers were produced, this had to be regarded as a significant disadvantage compared to well-defined Schrock initiators, which generally provide truly living polymerizations.

5.5
Ruthenium-Based Cyclopolymerization Systems

Grubbs-Herrmann (such as $RuCl_2$(1,3-bis(2,4,6-trimethylphenyl)-4,5-dihydroimidazolin-2-ylidene)(CHPh)(PCy$_3$)) and Grubbs-Hoveyda (such as $RuCl_2$(1,3-bis(2,4,6-trimethylphenyl)-4,5-dihydroimidazol-2-ylidene)(CH(2-(2-PrO-C_6H_4)) catalysts [132] are air and moisture-stable metathesis catalysts with remarkable activities, sometimes rivaling those of highly active Schrock catalysts [133]. Nonetheless, despite their activity in ring-opening metathesis polymerization (ROMP), ring-closing, enyne and ring-opening cross metathesis reactions, none of the systems examined so far have been capable of polymerizing alkynes or cyclopolymerizing 1,6-heptadiynes. Recently, Buchmeiser and Nuyken et al. reported on the successful synthesis of a modified Grubbs-Hoveyda catalyst that was capable of cyclopolymerizing 1,6-heptadiynes in a living and stereoregular way [134,135]. With the development of these catalystic systems, one of the last gaps between molybdenum- and ruthenium-based metathesis catalysts was closed.

Ru(CF$_3$COO)$_2$(1,3-dimesityl-4,5-dihydroimidazl-2-ylidene)(CH-(2-(2-PrO-C_6H_4))), initially developed as a highly efficient and active catalyst for ring-closing, enyne and cross metathesis [136], possesses enhanced polarization across the ruthenium-carbon double bond, which directly translates into increased reactivity, allowing the cyclopolymerization of DEDPM. Nevertheless, although highly active, this catalyst did not facilitate living polymerization [83] of DEDPM. Therefore, catalytic variations were tested to overcome this problem. Exchange of the 2-(2-propoxy)benzylidene ligand for the 2,4,5-trimethoxybenzylidene ligand [137] resulted in Ru(CF$_3$COO)$_2$(1,3-dimesityl-4,5-dihydroimidazol-2-ylidene)(CH-(2,4,5-(MeO)$_3$-C_6H_4)). This compound turned out to be an excellent catalyst for the cyclopolymerization of DEDPM, allowing full control over molecular weight [134]. Similar to polymerizations carried out with well-defined Schrock initiators, polymerizations of DEDPM with this catalyst in methylene chloride proceeded in a class VI living manner [84]. In terms of microstructure, poly(DEDPM) prepared by this catalyst again consisted virtually solely (>96%) of cyclopent-1-enylene-1-vinylene units, as shown by ^{13}C-NMR measurements [130, 131]. The polymerization mechanism was proposed to follow that for molybdenum-based cyclopolymerization [81] (Scheme 6), as evidenced by ^{1}H-NMR. Here, the disappearance of the starting alkylidene at δ=17.58 ppm and the appearance of new alkylidene signals at δ=15.63, 18.67, 19.90, 20.84 and 21.63 ppm was observed.

Using MALDI-TOF spectroscopy, 2-propoxybenzylidene was found to be the end group in all of the polymers prepared by the action of Ru(CF$_3$COO)$_2$ (1,3-dimesityl-4,5-dihydroimidazl-2-ylidene)(CH-(2-(2-PrO)-C_6H_4))), indicating the absence of any chain transfer reactions. The polymerization of chiral 4-(ethoxycarbonyl)-4-(1S, 2R, 5S)-(+)-menthoxycarbonyl-1,6-heptadiyne by this catalyst again proceeded in a stereo- and regioselective way. The ^{13}C-NMR spectrum of poly(4-(ethoxycarbonyl)-4-(1S, 2R, 5S)-(+)-menthoxycarbonyl-

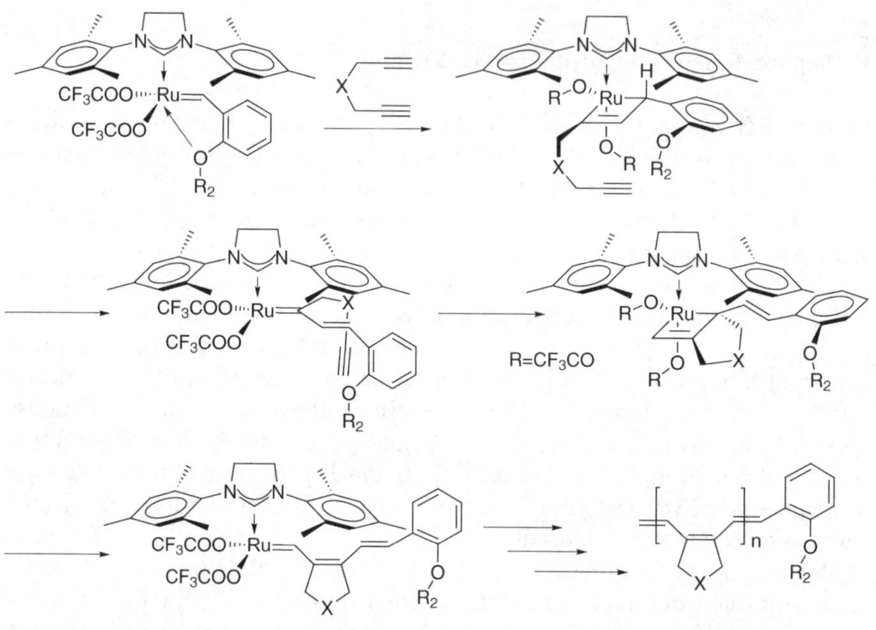

Scheme 6 Proposed mechanism of cyclopolymerization of 1,6-heptadiynes using a Ru-based initiator

1,6-heptadiyne) was identical to the one recorded with a sample of this polymer prepared by the action of a Schrock catalyst. Consequently, the same structure was assigned, a *st* poly(*trans*-cyclopent-1-enylene-1-vinylene) with >98% stereoregularity [130, 131]. This finding was of particular interest, since polymers prepared via ring-opening metathesis polymerization (ROMP) using other Ru-based catalysts have a *trans*-content ≤90% and low stereoregularity [133, 138].

In an extension to this work, Krause et al. varied the catalytic system systematically in order to further improve the entire polymerization system [135]. For this purpose, fourteen metathesis initiators – the Grubbs-Hoveyda catalyst $RuCl_2(IMesH_2)$ $(CH-2-(2-PrO)-C_6H_4)$, as well as $Ru(CF_3COO)_2(IMesH_2)(CH-2-(2-PrO)-C_6H_4)$, $Ru(CF_3CF_2COO)_2(IMesH_2)(CH-2-(2-PrO)-C_6H_4)$, $Ru(CF_3CF_2CF_2COO)_2(IMesH_2)(CH-2-(2-PrO)-C_6H_4)$, $RuCl_2(IMesH_2)(CH-2,4,5-(MeO)_3-C_6H_2)$, $Ru(CF_3COO)_2(IMesH_2)(CH-2,4,5-(MeO)_3-C_6H_2)$, $Ru(CF_3CF_2COO)_2(IMesH_2)(CH-2,4,5-(MeO)_3-C_6H_2)$, $Ru(CF_3CF_2CF_2COO)_2(IMesH_2)(CH-2,4,5-(MeO)_3-C_6H_2)$, $RuCl_2(IMes)$ $(CH-2-(2-PrO)-C_6H_4)$, $Ru(CF_3COO)_2(IMes)(CH-2-(2-PrO)-C_6H_4)$, $RuCl_2(IMesH_2)(CH-2-(2-PrO)-5-NO_2-C_6H_3)$, $Ru(CF_3COO)_2(IMesH_2)(CH-2-(2-PrO)-5-NO_2-C_6H_3)$, $Ru(CF_3CF_2COO)_2(IMesH_2)(CH-2-(2-PrO)-5-NO_2-C_6H_3)$, and $Ru(CF_3CF_2CF_2COO)_2(IMesH_2)(CH-2-(2-PrO)-5-NO_2-C_6H_3)$ ($IMes$=1,3-dimesitylimidazol-2-ylidene; $IMesH_2$=1,3-dimesityl-4,5-dihydroimidazol-2-ylidene, Fig. 11) – were investigated for their polymerization activity vs DEDPM.

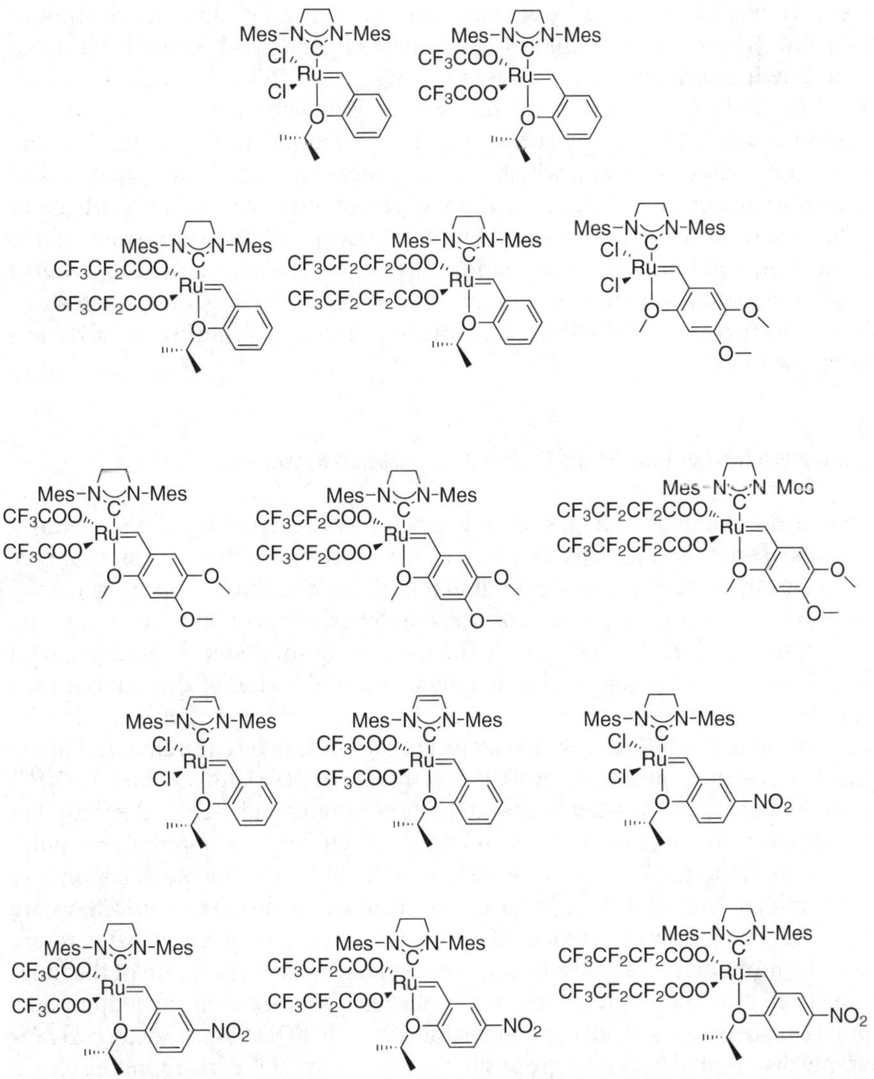

Fig. 11 Ru-based initiators used in a systematic study of the cyclopolymerization of DEDPM

Class VI living polymerization systems were generated with the initiators Ru(CF$_3$COO)$_2$(IMesH$_2$)(CH-2,4,5-(MeO)$_3$-C$_6$H$_2$), Ru(CF$_3$CF$_2$COO)$_2$(IMesH$_2$)(CH-2,4,5-(MeO)$_3$-C$_6$H$_2$), Ru(CF$_3$CF$_2$CF$_2$COO)$_2$(IMesH$_2$)(CH-2,4,5-(MeO)$_3$-C$_6$H$_2$), containing the (CH-2,4,5-(OMe)$_3$-C$_6$H$_2$) group, and the initiators Ru(CF$_3$COO)$_2$(IMesH$_2$)(CH-2-(2-PrO)-5-NO$_2$-C$_6$H$_3$), Ru(CF$_3$CF$_2$COO)$_2$(IMesH$_2$)(CH-2-(2-PrO)-5-NO$_2$-C$_6$H$_3$), and Ru(CF$_3$CF$_2$CF$_2$COO)$_2$(IMesH$_2$)(CH-2-(2-PrO)-5-NO$_2$-C$_6$H$_3$), containing the (CH-2-(2-PrO)-5-NO$_2$-C$_6$H$_3$) group. From these systematic variations in catalyst structure, the following conclusions were drawn. In order to be suitable for the living cyclopolymerization of DEDPM: (i) the replacement of both chlorine ligands with strongly electron-withdrawing carboxylic salts such as $CF_3(CF_2)_{x=0-2}COOAg$ is required; (ii) the NHC has to be electron rich. In addition, the living character of the polymerization of DEDPM strongly correlates with the steric and electronic situation at the benzylidene ligand, which directly influences, the values of k_p/k_i and therefore insertion efficiencies. Both types of initiators mentioned above fulfill these criteria. These initiators gave rise to 100% α-insertion of the monomer, resulting in the formation of poly(acetylene)s containing almost solely (>96%) five-membered ring structures. The use of larger fluorinated carboxylates further reduced the chain-transfer reactions, resulting in poly(ene)s with low PDIs.

5.6
Supported Ruthenium-Based Cyclopolymerization Systems

One of the many advantages of ruthenium-based catalysts is their tolerance to polar functional groups and water as a reaction medium. Consequently, reactions that could be carried out under aqueous conditions appear favorable. One of the ways to run polymerizations under aqueous conditions is to locate the catalyst within a micelle. While the use of sodium dodecylsulfonate (SDS) as micelle-forming reagent clearly failed, immobilization of the catalyst on a micelle-forming support was successful. The synthetic route necessary for the realization of a polymer-bound catalyst must fulfill two requirements. On the one hand, perfect mimics of $Ru(CF_3COO)_2(IMesH_2)(CH-(2,4,5-(MeO)_3-C_6H_4)$ must be generated in order to maintain its reactivity and stereoselectivity. On the other hand, and in contrast to conventional suspension/emulsion polymerization, the catalyst has to be permanently linked to the block copolymer amphiphile. The catalyst only locates itself in the hydrophobic micellar core upon micelle formation of the functionalized block copolymer in water, where also the monomer is dissolved, under these conditions. Preparation of the functionalized block copolymers was accomplished by first reacting an amphiphilc, poly(2-oxazoline)-derived copolymer, $Me_{30}Non_6(PenOH)_2Pip$, bearing two randomly distributed hydroxyl groups in the side chain of the hydrophobic block, with hexafluoroglutaric anhydride followed by deprotonation with aqueous $NaOH$ and reaction with $AgNO_3$ to yield a polymer-bound silver carboxylate. The final steps entailed its reaction with the catalyst precursors $RuCl_2$(1,3-bis(2,4,6-trimethylphenyl)-4,5-dihydroimidazol-2-ylidene)$(CH-2-(2-PrO)-C_6H_4$ and $RuCl_2$(1,3-bis(2,4,6-trimethylphenyl)-4,5-dihydroimidazol-2-ylidene) $(CH-2,4,5-(OMe)_3-C_6H_2)$ [137], respectively, followed by reaction with silver trifluoroacetate to endcap the unreacted chloro groups of the catalysts. In the course of this two-step chlorine exchange, the corresponding ruthenium compounds were fixed to the support to yield the poly(2-oxazoline)-immobilized catalyst $Me_{30}Non_6((PenOCO(CF_2)_3COO)(CF_3COO)Ru(CH-2,4,5-(OMe)_3C_6H_2)(IMesH_2))_{0.8}(PenOCO(CF_2)_3COOAg)_{1.2}Pip$ (Scheme 7).

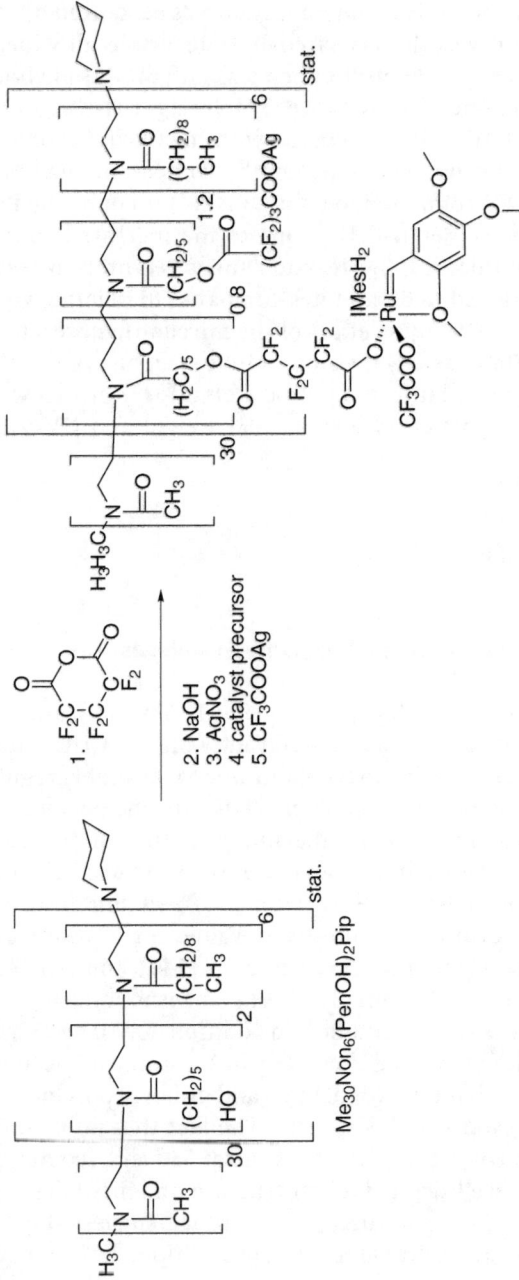

Scheme 7 Immobilization of RuCl$_2$(CH-2,4,5-(OMe)$_3$-C$_6$H$_2$)(1,3-bis(2,4,6-trimethylphenyl)-4,5-dihydroimidazol-2-ylidene) on a micelle-forming block-copolymer

The successful and selective immobilization was evidenced by ^{1}H-NMR. The immobilized system was characterized by one single alkylidene signal at δ=17.51 ppm, which perfectly fits the chemical shift for the alkylidene signal in the homologous compound Ru(CF$_3$COO)$_2$(IMesH$_2$)(CH-(2,4,5-(MeO)$_3$-C$_6$H$_4$) at δ=17.58 ppm. Poly(DEDPM) prepared with the block copolymer-immobilized initiator in water was characterized by comparably low polydispersity indices (PDIs) (<1.40) compared to poly(DEDPM) prepared by Ru(CF$_3$COO)$_2$ (IMesH$_2$)(CH-2-(2,4,5-(MeO)$_3$-C$_6$H$_3$). Due to the increased concentration of DEDPM within the micelles, the reaction times required to reach complete conversion were reduced to 30 minutes compared to 2 hours with the parent catalyst, indicating the catalytic effect of the micellar nanoreactors formed in aqueous medium. Stabilized by the amphiphilic structure of the block-copolymer, the poly(acetylene) lattices could be stored for over two weeks without any changes in latex particle size, molecular weight and UV-Vis absorption maxima.

5.7
Physical Properties of Poly(DEDPM)$_n$

5.7.1
Poly(DEDPM)$_n$ Based on Cyclopent-1-enylene-1-vinylenes [139]

Poly(DEDPM)$_n$ based on cyclopent-1-enylene-1-vinylene units is a dark-colored solid, that dissolves in chlorinated solvents (CH$_2$Cl$_2$, CHCl$_3$) to produce solutions with a deep violet color. Presumably due to the highly regular structure [130, 131], the polymers are insoluble in THF, benzene or toluene. The chain length had a strong influence on the fine structure of the corresponding UV-VIS spectra. Poly(DEDPM)$_5$ shows two absorption maxima, which become more pronounced in poly(DEDPM)$_n$ with a DP>10. For 10<n<50, a bathochromic shift in absorption was observed. Values for N_{eff} were calculated on the basis of poly(acetylene) model compounds [140], and were as high as 52, corresponding to a λ_{max} of 591 nm (THF). The corresponding molar absorption coefficient ε was 1.41×10^6 cm^2 mmol^{-1}. In addition, for 10<n<50, the intensity of λ_{max} increased with increasing chain length. According to the literature [141], only highly regular polymers with defined and uniform polymer architectures possess highly resolved UV-VIS spectra. The fact that poly(DEDPM)$_{70}$ possesses two well-resolved absorption maxima at 550 and 590 nm, respectively, was indicative of the well-defined microstructure, and fitted the structural data retrieved from ^{13}C-NMR spectroscopy well, which suggested a *st* alternating *cis-trans* structure (vide supra) [130, 131]. In addition, these poly(ene)s exhibited reversible thermochromic behaviors and excellent film-building properties. Poly(DEDPM)$_{10-90}$ exhibited glass transition temperatures (T_g) of around 26 °C, and was stable in air for months in the solid state as well as in solution (for example in CH$_2$Cl$_2$). Finally, poly(DEDPM)$_{10 \leq n \leq 90}$ was found to be thermally stable up to 185 °C under helium [130, 131].

The physical properties of other poly(1,6-heptadiynes) prepared by classic catalytic systems have been summarized elsewhere [103] and need to be treated carefully in view of the poor definitions of these materials.

5.7.2
Poly(DEDPM)$_n$ Based on Cylohex-1-ene-3-methylidenes [106]

Poly(DEDPM)$_n$≥50 consisting solely of six-membered rings displayed an absorption maximum λ_{max} of 511 nm, corresponding to a value for N_{eff} of 20 [24]. Not surprisingly at all, samples of poly(DEDPM) consisting of a mixture of five- and six-membered rings showed absorption maxima around 530 nm [81]. The fact that these poly(ene)s displayed reduced values for N_{eff} compared to their analogues based on cyclopent-1-enylene-1-vinylenes was believed to be a consequence of unfavorable sterics. In fact, copolymers of DEDPM with acetylene (1:1.5) displayed an absorption maximum at 598 nm [24], a value which is quite similar to the one reported for poly(DEDPM) based on five-membered rings (591 nm) [130, 131]. Since the acetylene monomer was shown to act as a spacer between single DEDPM-derived units [24], these findings strongly suggested that the comparably low absorption maximum of poly(DEDPM) based on six-membered rings stems from poor sterics. We should also mention that poly(DEDPM) based on cyclohex-1-ene-3-methylidene has been investigated recently for its non-linear optical properties [142].

6
Conclusions

Both 1-alkyne polymerization and cyclopolymerization have experienced dramatic improvements in terms of mechanistic understanding and definition of the resulting polymers and materials, respectively. The knowledge about complex catalytic systems has reached a level that allows the straightforward tailor-made synthesis of materials with designed properties. The physical properties (such as conductivity, photoconductivity) of the materials described here are currently under investigation, and many of the materials may have potential applications in, say, sensors. Despite these interesting and useful applications, often in the fields of physics or chemical engineering, one should not forget that any successful realization of "advanced materials" is dependent upon synthetic chemistry to a major extent. Furthermore, interdisciplinary research such as that described here, crossing the borders between organic, organometallic, physical, and polymer chemistry, is required.

Acknowledgement Our work was supported by the Austrian Science Fund (START Y-158). I wish to thank all undergraduate, graduate, and postgraduate students involved in the work described in this chapter for their dedication and enthusiasm.

References

1. Skotheim TA, Elsenbaumer RL, Reynolds JR (1997) Handbook of conducting polymers, 2nd edn. Marcel Dekker, New York
2. Brédas JL, Silbey RJ (1991) Conjugated polymers. Kluwer, Dordrecht
3. Heeger AJ (2001) Angew Chem 113:2660
4. MacDiarmid AG (2001) Angew Chem 113:2649
5. Shirakawa H (2001) Angew Chem 113:2642
6. Schottenberger H, Buchmeiser MR (1998) Recent Res Dev Macromol Res 3:535
7. Wagener KB, Boncella JM, Nel JG, Duttweiler RP, Hillmyer MA (1990) Makromol Chem 191:365
8. Wagener KB, Boncella JM, Nel JG (1991) Macromolecules 24:2649
9. Wolf A, Wagener KB (1991) Polym Prepr (Am Chem Soc, Div Polym Chem) 32:535
10. Thorn-Csányi E, Kraxner P (1997) Macromol Chem Phys 198: 3827
11. Thorn-Csányi E, Kraxner P (1997) J Mol Catal A 115:21
12. Thorn-Csányi E, Kraxner P (1997) Macromol Symp 122:77
13. Müllner R, Winkler B, Stelzer F, Tasch S, Hochfilzer C, Leising G (1999) Synthetic Met 105:129
14. Schlick H, Stelzer F, Tasch S, Leising G (2000) J Mol Catal A 160:71
15. Thorn-Csányi E, Herzog O (2004) J Mol Catal A 213:123
16. Brzezinska K, Wolfe PS, Watson MD, Wagener KB (1996) Macromol Chem Phys 197:2065
17. Krouse SA, Schrock RR (1988) Macromolecules 21:1885
18. Schlund R, Schrock RR, Crowe WE (1989) J Am Chem Soc 111:8004
19. Schaffer HE, Chance RR, Knoll K, Schrock RR, Silbey RJ (1990) Conjugated polymeric materials: Opportunities in electronics, optoelectronics, and molecular electronics. Kluwer, Dordrecht
20. Saunders RS, Cohen RE, Schrock RR (1991) Macromolecules 24:5599
21. Schrock RR, Luo S, Zanetti NC, Fox HH (1994) Organometallics 13:3396
22. Schrock RR (1995) Polyhedron 14:3177
23. Schrock RR, Luo S, Lee JC Jr, Zanetti NC, Davis WM (1996) J Am Chem Soc 118:3883
24. Schattenmann FJ, Schrock RR (1996) Macromolecules 29:8990
25. Abe Y, Masuda T, Higashimura T (1989) J Polym Sci Pol Chem 27:4267
26. Fujita Y, Misumi Y, Tabata M, Masuda T (1998) J Polym Sci Pol Chem 36:3157
27. Hayano S, Masuda T (1998) Macromolecules 31:3170
28. Hayano S, Masuda T (1999) Macromolecules 32:7344
29. Hayano S, Itoh T, Masuda T (1999) Polymer 40:4071
30. Masuda T, Hasegawa K, Higashimura T (1974) Macromolecules 7:728
31. Masuda T, Sasaki N, Higashimura T (1975) Macromolecules 8:717
32. Masuda T, Thieu K-Q, Sasaki N, Higashimura T (1976) Macromolecules 9:661
33. Masuda T, Higashimura T (1984) Acc Chem Res 17:51
34. Masuda T, Higashimura T (1986) Adv Polym Sci 81:121
35. Masuda T, Hamano T, Higashimura T, Ueda Tu, Muramatsu H (1988) Macromolecules 21:281
36. Masuda T, Hamano T, Tsuchihara K, Higashimura T (1990) Macromolecules 23:1374
37. Masuda T (1991) J Syn Org Chem Jpn 49:138
38. Masuda T, Mishima K, Fujimori J-I, Nishida M, Muramatsu H, Higashimura T (1992) Macromolecules 25:1401
39. Masuda T, Katahira S, Tsuchihara K, Higashimura T (1992) Polymer J 24:491
40. Masuda T, Fujimori JI, Abrahman MZ, Higashimura T (1993) Polymer J 25:535
41. Masuda T, Izumikawa H, Misumi Y, Higashimura T (1996) Macromolecules 29:1167

42. Masuda T, Kaneshiro H, Hayano S, Misumi Y, Bencze L (1997) J Macromol Sci Pure Appl Chem A34:1977
43. Masuda T, Hayano S, Iwawaki E, Nomura R (1998) J Mol Catal A 133:213
44. Masuda T, Abdul Karim SM, Nomura R (2000) J Mol Catal A 160:125
45. Minaki N, Hayano S, Masuda T (2002) Polymer 43:3579
46. Misumi Y, Masuda T (1998) Macromolecules 31:7572
47. Nakano M, Masuda T, Higashimura T (1994) Macromolecules 27:1344
48. Okano Y, Masuda T, Higashimura T (1987) J Polym Sci Pol Chem 25:1181
49. Sanda F, Kawaguchi T, Masuda T, Kobayashi N (2003) Macromolecules 36:2224
50. Sone T, R. Asako, Masuda T, Tabata M, Wada T, Sasabe H (2001) Macromolecules 34:1586
51. Tsuchihara K, Masuda T, Higashimura T, Nishida M, Muramatsu H (1990) Polym Bull 23:505
52. Yoshimura T, Masuda T, Higashimura T, Okuhara K, Ueda T (1991) Macromolecules 24:6053
53. Buchmeiser MR, Schrock RR (1995) Macromolecules 28:6642
54. Buchmeiser MR, Schuler N, Kaltenhauser G, Ongania K-H, Lagoja I, Wurst K, Schottenberger H (1998) Macromolecules 31:3175
55. Buchmeiser MR, Schuler N, Schottenberger H, Kohl I, Hallbrucker A (2000) Des Monomers Polym 3:421
56. Fijimori J, Masuda T, Higashimura T (1988) Polym Bull 20:1
57. Akiyoshi K, Masuda T, Higashimura T (1992) Macromol Chem Phys 193:755
58. Hirao T, Teramoto A, Sato T, Norisuye T, Masuda T, Higashimura T (1991) Polymer J 23:925
59. Isobe E, Masuda T, Higashimura T, Yamamoto A (1986) J Polym Sci Pol Chem 24:1839
60. Iwawaki E, Hayano S, Masuda T (2001) Polymer 42:4055
61. Izumikawa H, Masuda T, Higashimura T (1991) Polym Bull 27:193
62. Kouzai H, Masuda T, Higashimura T (1995) B Chem Soc Jpn 68:398
63. Masuda T, Isobe E, Higashimura T (1985) Macromolecules 18:841
64. Masuda T, Isobe E, Higashimura T (1983) J Am Chem Soc 105:7473
65. Masuda T, Yoshimura T, Higashimura T (1989) Macromolecules 22:3804
66. Matsumoto T, Masuda T, Higashimura T (1991) J Polym Sci Pol Chem 29:295
67. Tsuchihara K, Oshita T, Masuda T, Higashimura T (1991) Polymer J 23:1273
68. Yoshimura T, Masuda T, Higashimura T (1988) Macromolecules 21:1899
69. F. Hide, Díaz-García MA, Schwartz BJ, Heeger AJ (1997) Acc Chem Res 30:430
70. Sheats JR, Chang Y-L, Roitman DB, Stocking A (1999) Acc Chem Res 32:193
71. Fox MA (1999) Acc Chem Res 32:201
72. Horowitz G (1998) Adv Mater 10:365
73. Sirringhaus H, Tessler N, Friend RH (1998) Science 280:1741
74. Brédas J-L, Cornil J, Beljonne D, Santos DAD, Shuai Z (1999) Acc Chem Res 32:267
75. Hutten PF van, Krasnikov VV, Hadziioannou G (1999) Acc Chem Res 32:257
76. Wang YZ, Epstein AJ (1999) Acc Chem Rev 32: 217
77. Alivisatos AP, Barbara PF, Castlemann AW, Chang J, Dixon DA, Klein ML, McLendon GL, Miller JS, Ratner MA, Rossky PJ, Stupp SI, Thompson ME (1998) Adv Mater 10:1297
78. Chen C, Shi HJ, Tang CW (1997) Macromol Symp 125:1
79. Simionescu C, Lixandru T, Mazilu I, Tataru L (1971) Macromol Chem 147:69
80. Simionescu C, Lixandru T, Negulescu I, Mazilu I, Tataru L (1973) Makromol Chem 163:59
81. Fox HH, Schrock RR (1992) Organometallics 11:2763
82. Bazan GC, Khosravi E, Schrock RR, Feast JW, Gibson VC, O'Regan MB, Thomas JK, Davis WM (1990) J Am Chem Soc 112:8378

83. Darling TR, Davis TP, Fryd M, Gridnev AA, Haddleton DM, Ittel SD, Matheson RR Jr, Moad G, Rizzardo E (2000) J Polym Sci Pol Chem 38:1706
84. Matyjaszewski K (1993) Macromolecules 26:1787
85. Penczek S, Kubisa P, Szymanski R (1991) Makromol Chem Rapid Commun 12:77
86. Clemenson PI, Feast JW, Ahmad MM, Allen PC, Bott DC, Brown CS, Connors LM, Walker NS, Winter JN (1992) Polymer 33:4711
87. Dounis P, Feast JW, Widawski G (1997) J Mol Catal A 115:51
88. Saunders RS, Cohen RE, Schrock RR (1994) Acta Polym 45:301
89. Drury MR, Bloor D (1989) Synthetic Met 32:33
90. Buchmeiser M (1997) Macromolecules 30:2274
91. Shetty AS, Zhang J, Moore JS (1996) J Am Chem Soc 118:1019
92. Arnold R, Matchett SA, Rosenblum M (1988) Organometallics 7:2261
93. Schottenberger H, Buchmeiser MR, Herber R (2000) J. Organomet Chem 612:1
94. Fox HH, Wolf MO, O'Dell R, Lin BL, Schrock RR, Wrighton MS (1994) J Am Chem Soc 116:2827
95. Stille JK, Frey DA (1961) J Am Chem Soc 83:1697
96. Gibson HW, Epstein AJ, Rommelmann H, Tanner DB, Yang XQ, Pochan JM (1983) J Phys Colloq C3:651
97. Gibson HW, Bailey FC, Epstein AJ, Rommelmann H, Pochan JM (1980) Chem Commun 426
98. Gibson HW, Bailey FC, Epstein AJ, Rommelmann H, Kaplan S, Harbour J, Yang XQ, Tanner DB, Pochan JM (1983) J Am Chem Soc 105:4417
99. Aso C, Kunitake T, Saiki K (1972) Makromol Chem 151:265
100. Hubert AJ, Dale J (1990) J Chem Soc 3160
101. Harrell KJS, Nguyen ST (1999) Abstr Pap Am Chem S 217:121
102. Sivakumar C, Vasudevan T, Gopalan A, Wen T-C (2002) Polymer 43:1781
103. Choi S-K, Gal Y-S, Jin S-H, Kim H-K (2000) Chem Rev 100:1645
104. Colthup EC, Meriwether LS (1961) J Org Chem 26:5169
105. Gal YS, Jin SH, Choi SK (2004) J Mol Catal A 213:115
106. Schattenmann FJ, Schrock RR, Davis WM (1996) J Am Chem Soc 118:3295
107. Jang MS, Kwon SK, Choi SK (1990) Macromolecules 23:4135
108. Ryoo MS, Lee WC, Choi SK (1990) Macromolecules 23:3029
109. Kim YH, Choi KY, Choi SK (1989) J Polym Sci Pol Lett 27:443
110. Kim Y-H, Kwon S-K, Choi SK (1994) J Macromol Sci Pure Appl Chem A31:2041
111. Han SH, Kim UY, Kang YS, Choi S-K (1991) Macromolecules 24:973
112. Jin SH, H. N. Cho, Choi SK (1993) J Polym Sci Pol Chem 31:69
113. Lee H-J, Oh JM, Choi CJ, Kim H-K, Choi SK (1994) Polym Bull 32:433
114. Lee J-H, Park J-W, Oh J-M, Choi S-K (1995) Macromolecules 28:377
115. Kim Y-H, Kwon S-K, Lee J-K, Jeong KG, Choi SK (1995) J Macromol Sci Pure Appl Chem A32:1761
116. Kwon S-K, Kim Y-H, Choi S-K (1995) J Polym Sci Pol Chem 33:2135
117. Cho H-N, Lee J-Y, Kim S-H, Choi SK, Kim JY (1996) Polym Bull 36:391
118. Kim YH, Kwon SK, Choi SK (1997) Macromolecules 30:6677
119. Gal Y-S, Choi S-K (1995) Eur Polym J 31:941
120. Tlenkopatchev MA, Navarro J, Sanchev C, Canseco MA, Ogawa T (1995) Vysokomol Soedin 37:1212
121. Kang K, Cho LHN, Choi KY, Choi SK, Kim SH (1993) Macromolecules 26:4539
122. Kim S-H, S.-J. Choi, Park J-W, Cho H-N, Choi S-K (1994) Macromolecules 27:2339
123. Zhang N, Wu R, Li Q, Pakbaz K, Yoon CO, Wudl F (1993) Chem Mater 5:1598
124. Choi D-C, Kim S-H, Lee J-H, Cho H-N, Choi S-K (1997) Macromolecules 30:176

125. Buchmeiser MR (2000) Chem Rev 100:1565
126. Schrock RR, Lee J-K, O'Dell R, Oskam JH (1995) Macromolecules 28:5933
127. Schrock RR, Crowe WE, Bazan GC, M. DiMare, O'Regan MB, Schofield MH (1991) Organometallics 10:1832
128. Oskam JH, Schrock RR (1992) J Am Chem Soc 114:7588
129. Totland KM, Boyd TJ, Lavoie GG, Davis WM, Schrock RR (1996) Macromolecules 29:6114
130. Anders U, Nuyken O, Wurst K, Buchmeiser MR (2002) Angew Chem 114:4226
131. Anders U, Nuyken O, Wurst K, Buchmeiser MR (2002) Macromolecules 35:9029
132. Garber SB, Kingsbury JS, Gray BL, Hoveyda AH (2000) J Am Chem Soc 122: 8168
133. Bielawski CW, Grubbs RH (2000) Angew Chem 112:3025
134. Krause JO, Zarka MT, Anders U, Weberskirch R, Nuyken O, Buchmeiser MR (2003) Angew Chem 115:6147
135. Krause JO, Nuyken O, Buchmeiser MR (2004) Chem Eur J 10:2029
136. Krause JO, Wurst K, Nuyken O, Buchmeiser MR (2004) Chem Eur J 10:778
137. Grela K, Kim M (2003) Eur J Org Chem 963
138. Hamilton JG, Frenzel U, Kohl FJ, T. Weskamp, J. J. Rooney, W. A. Herrmann, Nuyken O (2000) J. Organomet Chem 606:8
139. Anders U, Nuyken O, Buchmeiser MR (2003) Des Monomer Polym 6:135
140. Knoll K, Schrock RR (1989) J Am Chem Soc 111:7989
141. Roncali J (1997) Chem Rev 97:173
142. Ledoux I, Samuel IDW, Zyss J, Yaliraki SN, Schattenmann FJ, Schrock RR, Silbey RJ (1999) Chem Phys 245:1

Received: May 2004

Author Index Volumes 101–176

Author Index Volumes 1–100 see Volume 100

de, Abajo, J. and *de la Campa, J. G.*: Processable Aromatic Polyimides. Vol. 140, pp. 23–60.
Abetz, V. see Förster, S.: Vol. 166, pp. 173–210.
Adolf, D. B. see Ediger, M. D.: Vol. 116, pp. 73–110.
Aharoni, S. M. and *Edwards, S. F.*: Rigid Polymer Networks. Vol. 118, pp. 1–231.
Albertsson, A.-C. and *Varma, I. K.*: Aliphatic Polyesters: Synthesis, Properties and Applications. Vol. 157, pp. 99–138.
Albertsson, A.-C. see Edlund, U.: Vol. 157, pp. 53–98.
Albertsson, A.-C. see Söderqvist Lindblad, M.: Vol. 157, pp. 139–161.
Albertsson, A.-C. see Stridsberg, K. M.: Vol. 157, pp. 27–51.
Albertsson, A.-C. see Al-Malaika, S.: Vol. 169, pp. 177–199.
Al-Malaika, S.: Perspectives in Stabilisation of Polyolefins. Vol. 169, pp. 121–150.
Améduri, B., Boutevin, B. and *Gramain, P.*: Synthesis of Block Copolymers by Radical Polymerization and Telomerization. Vol. 127, pp. 87–142.
Améduri, B. and *Boutevin, B.*: Synthesis and Properties of Fluorinated Telechelic Monodispersed Compounds. Vol. 102, pp. 133–170.
Amselem, S. see Domb, A. J.: Vol. 107, pp. 93–142.
Andrady, A. L.: Wavelenght Sensitivity in Polymer Photodegradation. Vol. 128, pp. 47–94.
Andreis, M. and *Koenig, J. L.*: Application of Nitrogen-15 NMR to Polymers. Vol. 124, pp. 191–238.
Angiolini, L. see Carlini, C.: Vol. 123, pp. 127–214.
Anjum, N. see Gupta, B.: Vol. 162, pp. 37–63.
Anseth, K. S., Newman, S. M. and *Bowman, C. N.*: Polymeric Dental Composites: Properties and Reaction Behavior of Multimethacrylate Dental Restorations. Vol. 122, pp. 177–218.
Antonietti, M. see Cölfen, H.: Vol. 150, pp. 67–187.
Armitage, B. A. see O'Brien, D. F.: Vol. 126, pp. 53–58.
Arndt, M. see Kaminski, W.: Vol. 127, pp. 143–187.
Arnold Jr., F. E. and *Arnold, F. E.*: Rigid-Rod Polymers and Molecular Composites. Vol. 117, pp. 257–296.
Arora, M. see Kumar, M. N. V. R.: Vol. 160, pp. 45–118.
Arshady, R.: Polymer Synthesis via Activated Esters: A New Dimension of Creativity in Macromolecular Chemistry. Vol. 111, pp. 1–42.
Auer, S. and *Frenkel, D.*: Numerical Simulation of Crystal Nucleation in Colloids. Vol. 173, pp. 149–208.

Bahar, I., Erman, B. and *Monnerie, L.*: Effect of Molecular Structure on Local Chain Dynamics: Analytical Approaches and Computational Methods. Vol. 116, pp. 145–206.
Ballauff, M. see Dingenouts, N.: Vol. 144, pp. 1–48.
Ballauff, M. see Holm, C.: Vol. 166, pp. 1–27.

Ballauff, M. see Rühe, J.: Vol. 165, pp. 79–150.
Baltá-Calleja, F. J., González Arche, A., Ezquerra, T. A., Santa Cruz, C., Batallón, F., Frick, B. and *López Cabarcos, E.*: Structure and Properties of Ferroelectric Copolymers of Poly(vinylidene) Fluoride. Vol. 108, pp. 1–48.
Barnes, M. D. see Otaigbe, J.U.: Vol. 154, pp. 1–86.
Barshtein, G. R. and *Sabsai, O. Y.*: Compositions with Mineralorganic Fillers.Vol. 101, pp. 1–28.
Barton, J. see Hunkeler, D.: Vol. 112, pp. 115–134.
Baschnagel, J., Binder, K., Doruker, P., Gusev, A. A., Hahn, O., Kremer, K., Mattice, W. L., Müller-Plathe, F., Murat, M., Paul, W., Santos, S., Sutter, U. W. and *Tries, V.*: Bridging the Gap Between Atomistic and Coarse-Grained Models of Polymers: Status and Perspectives. Vol. 152, pp. 41–156.
Batallán, F. see Baltá-Calleja, F. J.: Vol. 108, pp. 1–48.
Batog, A. E., Pet'ko, I. P. and *Penczek, P.*: Aliphatic-Cycloaliphatic Epoxy Compounds and Polymers. Vol. 144, pp. 49–114.
Baughman, T. W. and *Wagener, K. B.*: Recent Advances in ADMET Polymerization. Vol 176, pp. 1–42.
Bell, C. L. and *Peppas, N. A.*: Biomedical Membranes from Hydrogels and Interpolymer Complexes. Vol. 122, pp. 125–176.
Bellon-Maurel, A. see Calmon-Decriaud, A.: Vol. 135, pp. 207–226.
Bennett, D. E. see O'Brien, D. F.: Vol. 126, pp. 53–84.
Berry, G. C.: Static and Dynamic Light Scattering on Moderately Concentraded Solutions: Isotropic Solutions of Flexible and Rodlike Chains and Nematic Solutions of Rodlike Chains. Vol. 114, pp. 233–290.
Bershtein, V. A. and *Ryzhov, V. A.*: Far Infrared Spectroscopy of Polymers. Vol. 114, pp. 43–122.
Bhargava R., Wang S.-Q. and *Koenig J. L*: FTIR Microspectroscopy of Polymeric Systems. Vol. 163, pp. 137–191.
Biesalski, M.: see Rühe, J.: Vol. 165, pp. 79–150.
Bigg, D. M.: Thermal Conductivity of Heterophase Polymer Compositions.Vol. 119, pp. 1–30.
Binder, K.: Phase Transitions in Polymer Blends and Block Copolymer Melts: Some Recent Developments. Vol. 112, pp. 115–134.
Binder, K.: Phase Transitions of Polymer Blends and Block Copolymer Melts in Thin Films. Vol. 138, pp. 1–90.
Binder, K. see Baschnagel, J.: Vol. 152, pp. 41–156.
Binder, K., Müller, M., Virnau, P. and *González MacDowell, L.*: Polymer+Solvent Systems: Phase Diagrams, Interface Free Energies, and Nucleation. Vol. 173, pp. 1–104.
Bird, R. B. see Curtiss, C. F.: Vol. 125, pp. 1–102.
Biswas, M. and *Mukherjee, A.*: Synthesis and Evaluation of Metal-Containing Polymers. Vol. 115, pp. 89–124.
Biswas, M. and *Sinha Ray, S.*: Recent Progress in Synthesis and Evaluation of Polymer-Montmorillonite Nanocomposites. Vol. 155, pp. 167–221.
Bogdal, D., Penczek, P., Pielichowski, J. and *Prociak, A.*: Microwave Assisted Synthesis, Crosslinking, and Processing of Polymeric Materials. Vol. 163, pp. 193–263.
Bohrisch, J., Eisenbach, C.D., Jaeger, W., Mori H., Müller A.H.E., Rehahn, M., Schaller, C., Traser, S. and *Wittmeyer, P.*: New Polyelectrolyte Architectures. Vol. 165, pp. 1–41.
Bolze, J. see Dingenouts, N.: Vol. 144, pp. 1–48.
Bosshard, C.: see Gubler, U.: Vol. 158, pp. 123–190.
Boutevin, B. and *Robin, J. J.*: Synthesis and Properties of Fluorinated Diols. Vol. 102. pp. 105–132.

Boutevin, B. see Amédouri, B.: Vol. 102, pp. 133–170.
Boutevin, B. see Améduri, B.: Vol. 127, pp. 87–142.
Bowman, C. N. see Anseth, K. S.: Vol. 122, pp. 177–218.
Boyd, R. H.: Prediction of Polymer Crystal Structures and Properties. Vol. 116, pp. 1–26.
Briber, R. M. see Hedrick, J. L.: Vol. 141, pp. 1–44.
Bronnikov, S. V., Vettegren, V. I. and *Frenkel, S. Y.:* Kinetics of Deformation and Relaxation in Highly Oriented Polymers. Vol. 125, pp. 103–146.
Brown, H. R. see Creton, C.: Vol. 156, pp. 53–135.
Bruza, K. J. see Kirchhoff, R. A.: Vol. 117, pp. 1–66.
Buchmeiser, M. R.: Regioselective Polymerization of 1-Alkynes and Stereoselective Cyclopolymerization of α, ω-Heptadiynes. Vol. 176, pp. 89–119.
Budkowski, A.: Interfacial Phenomena in Thin Polymer Films: Phase Coexistence and Segregation. Vol. 148, pp. 1–112.
Burban, J. H. see Cussler, E. L.: Vol. 110, pp. 67–80.
Burchard,W.: Solution Properties of Branched Macromolecules. Vol. 143, pp. 113–194.
Butté, A. see Schork, F. J.: Vol. 175, pp. 129–255.

Calmon-Decriaud, A., Bellon-Maurel, V., Silvestre, F.: Standard Methods for Testing the Aerobic Biodegradation of Polymeric Materials.Vol 135, pp. 207–226.
Cameron, N. R. and *Sherrington, D. C.:* High Internal Phase Emulsions (HIPEs)-Structure, Properties and Use in Polymer Preparation.Vol. 126, pp. 163–214.
de la Campa, J. G. see de Abajo, J.: Vol. 140, pp. 23–60.
Candau, F. see Hunkeler, D.: Vol. 112, pp. 115–134.
Canelas, D. A. and *DeSimone, J. M.:* Polymerizations in Liquid and Supercritical Carbon Dioxide. Vol. 133, pp. 103–140.
Canva, M. and *Stegeman, G. I.:* Quadratic Parametric Interactions in Organic Waveguides. Vol. 158, pp. 87–121.
Capek, I.: Kinetics of the Free-Radical Emulsion Polymerization of Vinyl Chloride. Vol. 120, pp. 135–206.
Capek, I.: Radical Polymerization of Polyoxyethylene Macromonomers in Disperse Systems. Vol. 145, pp. 1–56.
Capek, I. and *Chern, C.-S.:* Radical Polymerization in Direct Mini-Emulsion Systems. Vol. 155, pp. 101–166.
Cappella, B. see Munz, M.: Vol. 164, pp. 87–210.
Carlesso, G. see Prokop, A.: Vol. 160, pp. 119–174.
Carlini, C. and *Angiolini, L.:* Polymers as Free Radical Photoinitiators. Vol. 123, pp. 127–214.
Carter, K. R. see Hedrick, J. L.: Vol. 141, pp. 1–44.
Casas-Vazquez, J. see Jou, D.: Vol. 120, pp. 207–266.
Chandrasekhar, V.: Polymer Solid Electrolytes: Synthesis and Structure. Vol 135, pp. 139–206.
Chang, J. Y. see Han, M. J.: Vol. 153, pp. 1–36.
Chang, T.: Recent Advances in Liquid Chromatography Analysis of Synthetic Polymers. Vol. 163, pp. 1–60.
Charleux, B. and *Faust R.:* Synthesis of Branched Polymers by Cationic Polymerization. Vol. 142, pp. 1–70.
Chen, P. see Jaffe, M.: Vol. 117, pp. 297–328.
Chern, C.-S. see Capek, I.: Vol. 155, pp. 101–166.
Chevolot, Y. see Mathieu, H. J.: Vol. 162, pp. 1–35.
Choe, E.-W. see Jaffe, M.: Vol. 117, pp. 297–328.

Chow, P. Y. and *Gan, L. M.*: Microemulsion Polymerizations and Reactions. Vol. 175, pp. 257–298.
Chow, T. S.: Glassy State Relaxation and Deformation in Polymers. Vol. 103, pp. 149–190.
Chujo, Y. see Uemura, T.: Vol. 167, pp. 81–106.
Chung, S.-J. see Lin, T.-C.: Vol. 161, pp. 157–193.
Chung, T.-S. see Jaffe, M.: Vol. 117, pp. 297–328.
Cölfen, H. and *Antonietti, M.*: Field-Flow Fractionation Techniques for Polymer and Colloid Analysis. Vol. 150, pp. 67–187.
Colmenero J. see Richter, D.: Vol. 174, in press
Comanita, B. see Roovers, J.: Vol. 142, pp. 179–228.
Connell, J. W. see Hergenrother, P. M.: Vol. 117, pp. 67–110.
Creton, C., Kramer, E. J., Brown, H. R. and *Hui, C.-Y.*: Adhesion and Fracture of Interfaces Between Immiscible Polymers: From the Molecular to the Continuum Scale. Vol. 156, pp. 53–135.
Criado-Sancho, M. see Jou, D.: Vol. 120, pp. 207–266.
Curro, J. G. see Schweizer, K. S.: Vol. 116, pp. 319–378.
Curtiss, C. F. and *Bird, R. B.*: Statistical Mechanics of Transport Phenomena: Polymeric Liquid Mixtures. Vol. 125, pp. 1–102.
Cussler, E. L., Wang, K. L. and *Burban, J. H.*: Hydrogels as Separation Agents. Vol. 110, pp. 67–80.

Dalton, L.: Nonlinear Optical Polymeric Materials: From Chromophore Design to Commercial Applications. Vol. 158, pp. 1–86.
Dautzenberg, H. see Holm, C.: Vol. 166, pp.113–171.
Davidson, J. M. see Prokop, A.: Vol. 160, pp.119–174.
Desai, S. M. and *Singh, R. P.*: Surface Modification of Polyethylene. Vol. 169, pp. 231–293.
DeSimone, J. M. see Canelas D. A.: Vol. 133, pp. 103–140.
DeSimone, J. M. see Kennedy, K. A.: Vol. 175, pp. 329–346.
DiMari, S. see Prokop, A.: Vol. 136, pp. 1–52.
Dimonie, M. V. see Hunkeler, D.: Vol. 112, pp. 115–134.
Dingenouts, N., Bolze, J., Pötschke, D. and *Ballauf, M.*: Analysis of Polymer Latexes by Small-Angle X-Ray Scattering. Vol. 144, pp. 1–48.
Dodd, L. R. and *Theodorou, D. N.*: Atomistic Monte Carlo Simulation and Continuum Mean Field Theory of the Structure and Equation of State Properties of Alkane and Polymer Melts. Vol. 116, pp. 249–282.
Doelker, E.: Cellulose Derivatives. Vol. 107, pp. 199–266.
Dolden, J. G.: Calculation of a Mesogenic Index with Emphasis Upon LC-Polyimides. Vol. 141, pp. 189 –245.
Domb, A. J., Amselem, S., Shah, J. and *Maniar, M.*: Polyanhydrides: Synthesis and Characterization. Vol. 107, pp. 93–142.
Domb, A. J. see Kumar, M. N. V. R.: Vol. 160, pp. 45118.
Doruker, P. see Baschnagel, J.: Vol. 152, pp. 41–156.
Dubois, P. see Mecerreyes, D.: Vol. 147, pp. 1–60.
Dubrovskii, S. A. see Kazanskii, K. S.: Vol. 104, pp. 97–134.
Dunkin, I. R. see Steinke, J.: Vol. 123, pp. 81–126.
Dunson, D. L. see McGrath, J. E.: Vol. 140, pp. 61–106.
Dziezok, P. see Rühe, J.: Vol. 165, pp. 79–150.

Eastmond, G. C.: Poly(ε-caprolactone) Blends. Vol. 149, pp. 59–223.
Economy, J. and *Goranov, K.*: Thermotropic Liquid Crystalline Polymers for High Performance Applications. Vol. 117, pp. 221–256.

Ediger, M. D. and *Adolf, D. B.*: Brownian Dynamics Simulations of Local Polymer Dynamics. Vol. 116, pp. 73–110.
Edlund, U. and *Albertsson, A.-C.*: Degradable Polymer Microspheres for Controlled Drug Delivery. Vol. 157, pp. 53–98.
Edwards, S. F. see Aharoni, S. M.: Vol. 118, pp. 1–231.
Eisenbach, C. D. see Bohrisch, J.: Vol. 165, pp. 1–41.
Endo, T. see Yagci, Y.: Vol. 127, pp. 59–86.
Engelhardt, H. and *Grosche, O.*: Capillary Electrophoresis in Polymer Analysis. Vol.150, pp. 189–217.
Engelhardt, H. and *Martin, H.*: Characterization of Synthetic Polyelectrolytes by Capillary Electrophoretic Methods. Vol. 165, pp. 211–247.
Eriksson, P. see Jacobson, K.: Vol. 169, pp. 151–176.
Erman, B. see Bahar, I.: Vol. 116, pp. 145–206.
Eschner, M. see Spange, S.: Vol. 165, pp. 43–78.
Estel, K. see Spange, S.: Vol. 165, pp. 43–78.
Ewen, B. and *Richter, D.*: Neutron Spin Echo Investigations on the Segmental Dynamics of Polymers in Melts, Networks and Solutions. Vol. 134, pp. 1–130.
Ezquerra, T. A. see Baltá-Calleja, F. J.: Vol. 108, pp. 1–48.

Fatkullin, N. see Kimmich, R.: Vol. 170, pp. 1–113.
Faust, R. see Charleux, B.: Vol. 142, pp. 1–70.
Faust, R. see Kwon, Y.: Vol. 167, pp. 107–135.
Fekete, E. see Pukánszky, B.: Vol. 139, pp. 109–154.
Fendler, J. H.: Membrane-Mimetic Approach to Advanced Materials. Vol. 113, pp. 1–209.
Fetters, L. J. see Xu, Z.: Vol. 120, pp. 1–50.
Fontenot, K. see Schork, F. J.: Vol. 175, pp. 129–255.
Förster, S., Abetz, V. and *Müller, A. H. E.*: Polyelectrolyte Block Copolymer Micelles. Vol. 166, pp. 173–210.
Förster, S. and *Schmidt, M.*: Polyelectrolytes in Solution. Vol. 120, pp. 51–134.
Freire, J. J.: Conformational Properties of Branched Polymers: Theory and Simulations. Vol. 143, pp. 35–112.
Frenkel, S. Y. see Bronnikov, S.V.: Vol. 125, pp. 103–146.
Frick, B. see Baltá-Calleja, F. J.: Vol. 108, pp. 1–48.
Fridman, M. L.: see Terent'eva, J. P.: Vol. 101, pp. 29–64.
Fuchs, G. see Trimmel, G.: Vol. 176, pp. 43–87.
Fukui, K. see Otaigbe, J. U.: Vol. 154, pp. 1–86.
Funke, W.: Microgels-Intramolecularly Crosslinked Macromolecules with a Globular Structure. Vol. 136, pp. 137–232.
Furusho, Y. see Takata, T.: Vol. 171, pp. 1–75.

Galina, H.: Mean-Field Kinetic Modeling of Polymerization: The Smoluchowski Coagulation Equation.Vol. 137, pp. 135–172.
Gan, L. M. see Chow, P. Y.: Vol. 175, pp. 257–298.
Ganesh, K. see Kishore, K.: Vol. 121, pp. 81–122.
Gaw, K. O. and *Kakimoto, M.*: Polyimide-Epoxy Composites. Vol. 140, pp. 107–136.
Geckeler, K. E. see Rivas, B.: Vol. 102, pp. 171–188.
Geckeler, K. E.: Soluble Polymer Supports for Liquid-Phase Synthesis. Vol. 121, pp. 31–80.
Gedde, U. W. and *Mattozzi, A.*: Polyethylene Morphology. Vol. 169, pp. 29–73.
Gehrke, S. H.: Synthesis, Equilibrium Swelling, Kinetics Permeability and Applications of Environmentally Responsive Gels. Vol. 110, pp. 81–144.

de Gennes, P.-G.: Flexible Polymers in Nanopores. Vol. 138, pp. 91–106.
Georgiou, S.: Laser Cleaning Methodologies of Polymer Substrates. Vol. 168, pp. 1–49.
Geuss, M. see Munz, M.: Vol. 164, pp. 87–210.
Giannelis, E. P., Krishnamoorti, R. and *Manias, E.*: Polymer-Silicate Nanocomposites: Model Systems for Confined Polymers and Polymer Brushes. Vol. 138, pp. 107–148.
Godovsky, D. Y.: Device Applications of Polymer-Nanocomposites. Vol. 153, pp. 163–205.
Godovsky, D. Y.: Electron Behavior and Magnetic Properties Polymer-Nanocomposites. Vol. 119, pp. 79–122.
González Arche, A. see Baltá-Calleja, F. J.: Vol. 108, pp. 1–48.
Goranov, K. see Economy, J.: Vol. 117, pp. 221–256.
Gramain, P. see Améduri, B.: Vol. 127, pp. 87–142.
Grest, G. S.: Normal and Shear Forces Between Polymer Brushes. Vol. 138, pp. 149–184.
Grigorescu, G. and *Kulicke, W.-M.*: Prediction of Viscoelastic Properties and Shear Stability of Polymers in Solution. Vol. 152, p. 1–40.
Gröhn, F. see Rühe, J.: Vol. 165, pp. 79–150.
Grosberg, A. and *Nechaev, S.*: Polymer Topology. Vol. 106, pp. 1–30.
Grosche, O. see Engelhardt, H.: Vol. 150, pp. 189–217.
Grubbs, R., Risse, W. and *Novac, B.*: The Development of Well-defined Catalysts for Ring-Opening Olefin Metathesis. Vol. 102, pp. 47–72.
Gubler, U. and *Bosshard, C.*: Molecular Design for Third-Order Nonlinear Optics. Vol. 158, pp. 123–190.
van Gunsteren, W. F. see Gusev, A. A.: Vol. 116, pp. 207–248.
Gupta, B., Anjum, N.: Plasma and Radiation-Induced Graft Modification of Polymers for Biomedical Applications. Vol. 162, pp. 37–63.
Gusev, A. A., Müller-Plathe, F., van Gunsteren, W. F. and *Suter, U. W.*: Dynamics of Small Molecules in Bulk Polymers. Vol. 116, pp. 207–248.
Gusev, A. A. see Baschnagel, J.: Vol. 152, pp. 41–156.
Guillot, J. see Hunkeler, D.: Vol. 112, pp. 115–134.
Guyot, A. and *Tauer, K.*: Reactive Surfactants in Emulsion Polymerization. Vol. 111, pp. 43–66.

Hadjichristidis, N., Pispas, S., Pitsikalis, M., Iatrou, H. and *Vlahos, C.*: Asymmetric Star Polymers Synthesis and Properties. Vol. 142, pp. 71–128.
Hadjichristidis, N. see Xu, Z.: Vol. 120, pp. 1–50.
Hadjichristidis, N. see Pitsikalis, M.: Vol. 135, pp. 1–138.
Hahn, O. see Baschnagel, J.: Vol. 152, pp. 41–156.
Hakkarainen, M.: Aliphatic Polyesters: Abiotic and Biotic Degradation and Degradation Products. Vol. 157, pp. 1–26.
Hakkarainen, M. and *Albertsson, A.-C.*: Environmental Degradation of Polyethylene. Vol. 169, pp. 177–199.
Hall, H. K. see Penelle, J.: Vol. 102, pp. 73–104.
Hamley, I.W.: Crystallization in Block Copolymers. Vol. 148, pp. 113–138.
Hammouda, B.: SANS from Homogeneous Polymer Mixtures: A Unified Overview. Vol. 106, pp. 87–134.
Han, M. J. and *Chang, J. Y.*: Polynucleotide Analogues. Vol. 153, pp. 1–36.
Harada, A.: Design and Construction of Supramolecular Architectures Consisting of Cyclodextrins and Polymers. Vol. 133, pp. 141–192.
Haralson, M. A. see Prokop, A.: Vol. 136, pp. 1–52.
Hassan, C. M. and *Peppas, N. A.*: Structure and Applications of Poly(vinyl alcohol) Hydrogels Produced by Conventional Crosslinking or by Freezing/Thawing Methods. Vol. 153, pp. 37–65.

Hawker, C. J.: Dentritic and Hyperbranched Macromolecules Precisely Controlled Macromolecular Architectures. Vol. 147, pp. 113–160.
Hawker, C. J. see *Hedrick, J. L.*: Vol. 141, pp. 1–44.
He, G. S. see *Lin, T.-C.*: Vol. 161, pp. 157–193.
Hedrick, J. L., Carter, K. R., Labadie, J. W., Miller, R. D., Volksen, W., Hawker, C. J., Yoon, D. Y., Russell, T. P., McGrath, J. E. and *Briber, R. M.*: Nanoporous Polyimides. Vol. 141, pp. 1–44.
Hedrick, J. L., Labadie, J. W., Volksen, W. and *Hilborn, J. G.*: Nanoscopically Engineered Polyimides. Vol. 147, pp. 61–112.
Hedrick, J. L. see *Hergenrother, P. M.*: Vol. 117, pp. 67–110.
Hedrick, J. L. see *Kiefer, J.*: Vol. 147, pp. 161–247.
Hedrick, J. L. see *McGrath, J. E.*: Vol. 140, pp. 61–106.
Heine, D. R., Grest, G. S. and *Curro, J. G.*: Structure of Polymer Melts and Blends: Comparison of Integral Equation theory and Computer Sumulation. Vol. 173, pp. 209–249.
Heinrich, G. and *Klüppel, M.*: Recent Advances in the Theory of Filler Networking in Elastomers. Vol. 160, pp. 1–44.
Heller, J.: Poly (Ortho Esters). Vol. 107, pp. 41–92.
Helm, C. A.: see *Möhwald, H.*: Vol. 165, pp. 151–175.
Hemielec, A. A. see *Hunkeler, D.*: Vol. 112, pp. 115–134.
Hergenrother, P. M., Connell, J. W., Labadie, J. W. and *Hedrick, J. L.*: Poly(arylene ether)s Containing Heterocyclic Units. Vol. 117, pp. 67–110.
Hernández-Barajas, J. see *Wandrey, C.*: Vol. 145, pp. 123–182.
Hervet, H. see *Léger, L.*: Vol. 138, pp. 185–226.
Hilborn, J. G. see *Hedrick, J. L.*: Vol. 147, pp. 61–112.
Hilborn, J. G. see *Kiefer, J.*: Vol. 147, pp. 161–247.
Hiramatsu, N. see *Matsushige, M.*: Vol. 125, pp. 147–186.
Hirasa, O. see *Suzuki, M.*: Vol. 110, pp. 241–262.
Hirotsu, S.: Coexistence of Phases and the Nature of First-Order Transition in Poly-N-isopropylacrylamide Gels. Vol. 110, pp. 1–26.
Höcker, H. see *Klee, D.*: Vol. 149, pp. 1–57.
Holm, C., Hofmann, T., Joanny, J. F., Kremer, K., Netz, R. R., Reineker, P., Seidel, C., Vilgis, T. A. and *Winkler, R. G.: Polyelectrolyte Theory.* Vol. 166, pp. 67–111.
Holm, C., Rehahn, M., Oppermann, W. and *Ballauff, M.*: Stiff-Chain Polyelectrolytes. Vol. 166, pp. 1–27.
Hornsby, P.: Rheology, Compounding and Processing of Filled Thermoplastics. Vol. 139, pp. 155–216.
Houbenov, N. see *Rühe, J.*: Vol. 165, pp. 79–150.
Huber, K. see *Volk, N.*: Vol. 166, pp. 29–65.
Hugenberg, N. see *Rühe, J.*: Vol. 165, pp. 79–150.
Hui, C.-Y. see *Creton, C.*: Vol. 156, pp. 53–135.
Hult, A., Johansson, M. and *Malmström, E.*: Hyperbranched Polymers. Vol. 143, pp. 1–34.
Hünenberger, P. H.: Thermostat Algorithms for Molecular-Dynamics Simulations. Vol. 173, pp. 105–147.
Hunkeler, D., Candau, F., Pichot, C., Hemielec, A. E., Xie, T. Y., Barton, J., Vaskova, V., Guillot, J., Dimonie, M. V. and *Reichert, K. H.*: Heterophase Polymerization: A Physical and Kinetic Comparision and Categorization. Vol. 112, pp. 115–134.
Hunkeler, D. see *Macko, T.*: Vol. 163, pp. 61–136.
Hunkeler, D. see *Prokop, A.*: Vol. 136, pp. 1–52; 53–74.
Hunkeler, D. see *Wandrey, C.*: Vol. 145, pp. 123–182.

Iatrou, H. see *Hadjichristidis, N.*: Vol. 142, pp. 71–128.
Ichikawa, T. see *Yoshida, H.*: Vol. 105, pp. 3–36.

Ihara, E. see Yasuda, H.: Vol. 133, pp. 53–102.
Ikada, Y. see Uyama,Y.: Vol. 137, pp. 1–40.
Ikehara, T. see Jinnuai, H.: Vol. 170, pp. 115–167.
Ilavsky, M.: Effect on Phase Transition on Swelling and Mechanical Behavior of Synthetic Hydrogels. Vol. 109, pp. 173–206.
Imai, Y.: Rapid Synthesis of Polyimides from Nylon-Salt Monomers. Vol. 140, pp. 1–23.
Inomata, H. see Saito, S.: Vol. 106, pp. 207–232.
Inoue, S. see Sugimoto, H.: Vol. 146, pp. 39–120.
Irie, M.: Stimuli-Responsive Poly(N-isopropylacrylamide), Photo- and Chemical-Induced Phase Transitions. Vol. 110, pp. 49–66.
Ise, N. see Matsuoka, H.: Vol. 114, pp. 187–232.
Ito, H.: Chemical Amplification Resists for Microlithography. Vol. 172, pp. 37–245.
Ito, K. and *Kawaguchi, S.*: Poly(macronomers), Homo- and Copolymerization. Vol. 142, pp. 129–178.
Ito, K. see Kawaguchi, S.: Vol. 175, pp. 299–328.
Ito, Y. see Suginome, M.: Vol. 171, pp. 77–136.
Ivanov, A. E. see Zubov, V. P.: Vol. 104, pp. 135–176.

Jacob, S. and *Kennedy, J.*: Synthesis, Characterization and Properties of OCTA-ARM Polyisobutylene-Based Star Polymers. Vol. 146, pp. 1–38.
Jacobson, K., Eriksson, P., Reitberger, T. and *Stenberg, B.*: Chemiluminescence as a Tool for Polyolefin. Vol. 169, pp. 151–176.
Jaeger, W. see Bohrisch, J.: Vol. 165, pp. 1–41.
Jaffe, M., Chen, P., Choe, E.-W., Chung, T.-S. and *Makhija, S.*: High Performance Polymer Blends. Vol. 117, pp. 297–328.
Jancar, J.: Structure-Property Relationships in Thermoplastic Matrices. Vol. 139, pp. 1–66.
Jen, A. K.-Y. see Kajzar, F.: Vol. 161, pp. 1–85.
Jerome, R. see Mecerreyes, D.: Vol. 147, pp. 1–60.
Jiang, M., Li, M., Xiang, M. and *Zhou, H.*: Interpolymer Complexation and Miscibility and Enhancement by Hydrogen Bonding. Vol. 146, pp. 121–194.
Jin, J. see Shim, H.-K.: Vol. 158, pp. 191–241.
Jinnai, H., Nishikawa, Y., Ikehara, T. and *Nishi, T.*: Emerging Technologies for the 3D Analysis of Polymer Structures. Vol. 170, pp. 115–167.
Jo, W. H. and *Yang, J. S.*: Molecular Simulation Approaches for Multiphase Polymer Systems. Vol. 156, pp. 1–52.
Joanny, J.-F. see Holm, C.: Vol. 166, pp. 67–111.
Joanny, J.-F. see Thünemann, A. F.: Vol. 166, pp. 113–171.
Johannsmann, D. see Rühe, J.: Vol. 165, pp. 79–150.
Johansson, M. see Hult, A.: Vol. 143, pp. 1–34.
Joos-Müller, B. see Funke, W.: Vol. 136, pp. 137–232.
Jou, D., Casas-Vazquez, J. and *Criado-Sancho, M.*: Thermodynamics of Polymer Solutions under Flow: Phase Separation and Polymer Degradation. Vol. 120, pp. 207–266.

Kaetsu, I.: Radiation Synthesis of Polymeric Materials for Biomedical and Biochemical Applications. Vol. 105, pp. 81–98.
Kaji, K. see Kanaya, T.: Vol. 154, pp. 87–141.
Kajzar, F., Lee, K.-S. and *Jen, A. K.-Y.*: Polymeric Materials and their Orientation Techniques for Second-Order Nonlinear Optics. Vol. 161, pp. 1–85.
Kakimoto, M. see Gaw, K. O.: Vol. 140, pp. 107–136.
Kaminski, W. and *Arndt, M.*: Metallocenes for Polymer Catalysis. Vol. 127, pp. 143–187.

Kammer, H. W., Kressler, H. and *Kummerloewe, C.*: Phase Behavior of Polymer Blends – Effects of Thermodynamics and Rheology. Vol. 106, pp. 31–86.
Kanaya, T. and *Kaji, K.*: Dynamcis in the Glassy State and Near the Glass Transition of Amorphous Polymers as Studied by Neutron Scattering. Vol. 154, pp. 87–141.
Kandyrin, L. B. and *Kuleznev, V. N.*: The Dependence of Viscosity on the Composition of Concentrated Dispersions and the Free Volume Concept of Disperse Systems. Vol. 103, pp. 103–148.
Kaneko, M. see Ramaraj, R.: Vol. 123, pp. 215–242.
Kang, E. T., Neoh, K. G. and *Tan, K. L.*: X-Ray Photoelectron Spectroscopic Studies of Electroactive Polymers. Vol. 106, pp. 135–190.
Karlsson, S. see Söderqvist Lindblad, M.: Vol. 157, pp. 139–161.
Karlsson, S.: Recycled Polyolefins. Material Properties and Means for Quality Determination. Vol. 169, pp. 201–229.
Kato, K. see Uyama,Y.: Vol. 137, pp. 1–40.
Kautek, W. see Krüger, J.: Vol. 168, pp. 247–290.
Kawaguchi, S. see Ito, K.: Vol. 142, p 129–178.
Kawaguchi, S. and *Ito, K.*: Dispersion Polymerization. Vol. 175, pp. 299–328.
Kawata, S. see Sun, H.-B.: Vol. 170, pp. 169–273.
Kazanskii, K. S. and *Dubrovskii, S. A.*: Chemistry and Physics of Agricultural Hydrogels. Vol. 104, pp. 97–134.
Kennedy, J. P. see Jacob, S.: Vol. 146, pp. 1–38.
Kennedy, J. P. see Majoros, I.: Vol. 112, pp. 1–113.
Kennedy, K. A., Roberts, G. W. and *DeSimone, J. M.*: Heterogeneous Polymerization of Fluoroolefins in Supercritical Carbon Dioxide. Vol. 175, pp. 329–346.
Khokhlov, A., Starodybtzev, S. and *Vasilevskaya, V.*: Conformational Transitions of Polymer Gels: Theory and Experiment. Vol. 109, pp. 121–172.
Kiefer, J., Hedrick J. L. and *Hiborn, J. G.*: Macroporous Thermosets by Chemically Induced Phase Separation. Vol. 147, pp. 161–247.
Kihara, N. see Takata, T.: Vol. 171, pp. 1–75.
Kilian, H. G. and *Pieper, T.*: Packing of Chain Segments. A Method for Describing X-Ray Patterns of Crystalline, Liquid Crystalline and Non-Crystalline Polymers. Vol. 108, pp. 49–90.
Kim, J. see Quirk, R.P.: Vol. 153, pp. 67–162.
Kim, K.-S. see Lin, T.-C.: Vol. 161, pp. 157–193.
Kimmich, R. and *Fatkullin, N.*: Polymer Chain Dynamics and NMR. Vol. 170, pp. 1–113.
Kippelen, B. and *Peyghambarian, N.*: Photorefractive Polymers and their Applications. Vol. 161, pp. 87–156.
Kirchhoff, R. A. and *Bruza, K. J.*: Polymers from Benzocyclobutenes. Vol. 117, pp. 1–66.
Kishore, K. and *Ganesh, K.*: Polymers Containing Disulfide, Tetrasulfide, Diselenide and Ditelluride Linkages in the Main Chain. Vol. 121, pp. 81–122.
Kitamaru, R.: Phase Structure of Polyethylene and Other Crystalline Polymers by Solid-State 13C/MNR. Vol. 137, pp. 41–102.
Klee, D. and *Höcker, H.*: Polymers for Biomedical Applications: Improvement of the Interface Compatibility. Vol. 149, pp. 1–57.
Klier, J. see Scranton, A. B.: Vol. 122, pp. 1–54.
v. Klitzing, R. and *Tieke, B.*: Polyelectrolyte Membranes. Vol. 165, pp. 177–210.
Klüppel, M.: The Role of Disorder in Filler Reinforcement of Elastomers on Various Length Scales. Vol. 164, pp. 1–86.
Klüppel, M. see Heinrich, G.: Vol. 160, pp. 1–44.
Knuuttila, H., Lehtinen, A. and *Nummila-Pakarinen, A.*: Advanced Polyethylene Technologies – Controlled Material Properties. Vol. 169, pp. 13–27.

Kobayashi, S., Shoda, S. and *Uyama, H.:* Enzymatic Polymerization and Oligomerization. Vol. 121, pp. 1–30.
Köhler, W. and *Schäfer, R.:* Polymer Analysis by Thermal-Diffusion Forced Rayleigh Scattering. Vol. 151, pp. 1–59.
Koenig, J. L. see Bhargava, R.: Vol. 163, pp. 137–191.
Koenig, J. L. see Andreis, M.: Vol. 124, pp. 191–238.
Koike, T.: Viscoelastic Behavior of Epoxy Resins Before Crosslinking. Vol. 148, pp. 139–188.
Kokko, E. see Löfgren, B.: Vol. 169, pp. 1–12.
Kokufuta, E.: Novel Applications for Stimulus-Sensitive Polymer Gels in the Preparation of Functional Immobilized Biocatalysts. Vol. 110, pp. 157–178.
Konno, M. see Saito, S.: Vol. 109, pp. 207–232.
Konradi, R. see Rühe, J.: Vol. 165, pp. 79–150.
Kopecek, J. see Putnam, D.: Vol. 122, pp. 55–124.
Koßmehl, G. see Schopf, G.: Vol. 129, pp. 1–145.
Kozlov, E. see Prokop, A.: Vol. 160, pp. 119–174.
Kramer, E. J. see Creton, C.: Vol. 156, pp. 53–135.
Kremer, K. see Baschnagel, J.: Vol. 152, pp. 41–156.
Kremer, K. see Holm, C.: Vol. 166, pp. 67–111.
Kressler, J. see Kammer, H. W.: Vol. 106, pp. 31–86.
Kricheldorf, H. R.: Liquid-Cristalline Polyimides. Vol. 141, pp. 83–188.
Krishnamoorti, R. see Giannelis, E. P.: Vol. 138, pp. 107–148.
Krüger, J. and *Kautek, W.:* Ultrashort Pulse Laser Interaction with Dielectrics and Polymers, Vol. 168, pp. 247–290.
Kuchanov, S. I.: Modern Aspects of Quantitative Theory of Free-Radical Copolymerization. Vol. 103, pp. 1–102.
Kuchanov, S. I.: Principles of Quantitive Description of Chemical Structure of Synthetic Polymers. Vol. 152, p. 157–202.
Kudaibergennow, S. E.: Recent Advances in Studying of Synthetic Polyampholytes in Solutions. Vol. 144, pp. 115–198.
Kuleznev, V. N. see Kandyrin, L. B.: Vol. 103, pp. 103–148.
Kulichkhin, S. G. see Malkin, A. Y.: Vol. 101, pp. 217–258.
Kulicke, W.-M. see Grigorescu, G.: Vol. 152, p. 1–40.
Kumar, M. N. V. R., Kumar, N., Domb, A. J. and *Arora, M.:* Pharmaceutical Polymeric Controlled Drug Delivery Systems. Vol. 160, pp. 45–118.
Kumar, N. see Kumar M. N. V. R.: Vol. 160, pp. 45–118.
Kummerloewe, C. see Kammer, H. W.: Vol. 106, pp. 31–86.
Kuznetsova, N. P. see Samsonov, G.V.: Vol. 104, pp. 1–50.
Kwon, Y. and *Faust, R.:* Synthesis of Polyisobutylene-Based Block Copolymers with Precisely Controlled Architecture by Living Cationic Polymerization. Vol. 167, pp. 107–135.

Labadie, J. W. see Hergenrother, P. M.: Vol. 117, pp. 67–110.
Labadie, J. W. see Hedrick, J. L.: Vol. 141, pp. 1–44.
Labadie, J. W. see Hedrick, J. L.: Vol. 147, pp. 61–112.
Lamparski, H. G. see O'Brien, D. F.: Vol. 126, pp. 53–84.
Laschewsky, A.: Molecular Concepts, Self-Organisation and Properties of Polysoaps. Vol. 124, pp. 1–86.
Laso, M. see Leontidis, E.: Vol. 116, pp. 283–318.
Lazár, M. and *Rychl, R.:* Oxidation of Hydrocarbon Polymers. Vol. 102, pp. 189–222.
Lechowicz, J. see Galina, H.: Vol. 137, pp. 135–172.

Léger, L., Raphaël, E. and *Hervet, H.*: Surface-Anchored Polymer Chains: Their Role in Adhesion and Friction. Vol. 138, pp. 185–226.
Lenz, R. W.: Biodegradable Polymers. Vol. 107, pp. 1–40.
Leontidis, E., de Pablo, J. J., Laso, M. and *Suter, U. W.*: A Critical Evaluation of Novel Algorithms for the Off-Lattice Monte Carlo Simulation of Condensed Polymer Phases. Vol. 116, pp. 283–318.
Lee, B. see Quirk, R. P.: Vol. 153, pp. 67–162.
Lee, K.-S. see Kajzar, F.: Vol. 161, pp. 1–85.
Lee, Y. see Quirk, R. P: Vol. 153, pp. 67–162.
Lehtinen, A. see Knuuttila, H.: Vol. 169, pp. 13–27.
Leónard, D. see Mathieu, H. J.: Vol. 162, pp. 1–35.
Lesec, J. see Viovy, J.-L.: Vol. 114, pp. 1–42.
Li, M. see Jiang, M.: Vol. 146, pp. 121–194.
Liang, G. L. see Sumpter, B. G.: Vol. 116, pp. 27–72.
Lienert, K.-W.: Poly(ester-imide)s for Industrial Use. Vol. 141, pp. 45–82.
Lin, J. and *Sherrington, D. C.*: Recent Developments in the Synthesis, Thermostability and Liquid Crystal Properties of Aromatic Polyamides. Vol. 111, pp. 177–220.
Lin, T.-C., Chung, S.-J., Kim, K.-S., Wang, X., He, G. S., Swiatkiewicz, J., Pudavar, H. E. and *Prasad, P. N.*: Organics and Polymers with High Two-Photon Activities and their Applications. Vol. 161, pp. 157–193.
Lippert, T.: Laser Application of Polymers. Vol. 168, pp. 51–246.
Liu, Y. see Söderqvist Lindblad, M.: Vol. 157, pp. 139–161.
López Cabarcos, E. see Baltá-Calleja, F. J.: Vol. 108, pp. 1–48.
Löfgren, B., Kokko, E. and *Seppälä, J.*: Specific Structures Enabled by Metallocene Catalysis in Polyethenes. Vol. 169, pp. 1–12.
Löwen, H. see Thünemann, A. F.: Vol. 166, pp. 113–171.
Luo, Y. see Schork, F. J.: Vol. 175, pp. 129–255.

Macko, T. and *Hunkeler, D.*: Liquid Chromatography under Critical and Limiting Conditions: A Survey of Experimental Systems for Synthetic Polymers. Vol. 163, pp. 61–136.
Majoros, I., Nagy, A. and *Kennedy, J. P.*: Conventional and Living Carbocationic Polymerizations United. I. A Comprehensive Model and New Diagnostic Method to Probe the Mechanism of Homopolymerizations. Vol. 112, pp. 1–113.
Makhija, S. see Jaffe, M.: Vol. 117, pp. 297–328.
Malmström, E. see Hult, A.: Vol. 143, pp. 1–34.
Malkin, A. Y. and *Kulichkhin, S. G.*: Rheokinetics of Curing. Vol. 101, pp. 217–258.
Maniar, M. see Domb, A. J.: Vol. 107, pp. 93–142.
Manias, E. see Giannelis, E. P.: Vol. 138, pp. 107–148.
Martin, H. see Engelhardt, H.: Vol. 165, pp. 211–247.
Marty, J. D. and *Mauzac, M.*: Molecular Imprinting: State of the Art and Perspectives. Vol. 172, pp. 1–35.
Mashima, K., Nakayama, Y. and *Nakamura, A.*: Recent Trends in Polymerization of a-Olefins Catalyzed by Organometallic Complexes of Early Transition Metals.Vol. 133, pp. 1–52.
Mathew, D. see Reghunadhan Nair, C.P.: Vol. 155, pp. 1–99.
Mathieu, H. J., Chevolot, Y, Ruiz-Taylor, L. and *Leónard, D.*: Engineering and Characterization of Polymer Surfaces for Biomedical Applications. Vol. 162, pp. 1–35.
Matsumoto, A.: Free-Radical Crosslinking Polymerization and Copolymerization of Multivinyl Compounds. Vol. 123, pp. 41–80.
Matsumoto, A. see Otsu, T.: Vol. 136, pp. 75–138.

Matsuoka, H. and *Ise, N.*: Small-Angle and Ultra-Small Angle Scattering Study of the Ordered Structure in Polyelectrolyte Solutions and Colloidal Dispersions. Vol. 114, pp. 187–232.
Matsushige, K., Hiramatsu, N. and *Okabe, H.*: Ultrasonic Spectroscopy for Polymeric Materials. Vol. 125, pp. 147–186.
Mattice, W. L. see Rehahn, M.: Vol. 131/132, pp. 1–475.
Mattice, W. L. see Baschnagel, J.: Vol. 152, pp. 41–156.
Mattozzi, A. see Gedde, U. W.: Vol. 169, pp. 29–73.
Mauzac, M. see Marty, J. D.: Vol. 172, pp. 1–35.
Mays, W. see Xu, Z.: Vol. 120, pp. 1–50.
Mays, J. W. see Pitsikalis, M.: Vol. 135, pp. 1–138.
McGrath, J. E. see Hedrick, J. L.: Vol. 141, pp. 1–44.
McGrath, J. E., Dunson, D. L. and *Hedrick, J. L.*: Synthesis and Characterization of Segmented Polyimide-Polyorganosiloxane Copolymers. Vol. 140, pp. 61–106.
McLeish, T. C. B. and *Milner, S. T.*: Entangled Dynamics and Melt Flow of Branched Polymers. Vol. 143, pp. 195–256.
Mecerreyes, D., Dubois, P. and *Jerome, R.*: Novel Macromolecular Architectures Based on Aliphatic Polyesters: Relevance of the Coordination-Insertion Ring-Opening Polymerization. Vol. 147, pp. 1–60.
Mecham, S. J. see McGrath, J. E.: Vol. 140, pp. 61–106.
Menzel, H. see Möhwald, H.: Vol. 165, pp. 151–175.
Meyer, T. see Spange, S.: Vol. 165, pp. 43–78.
Mikos, A. G. see Thomson, R. C.: Vol. 122, pp. 245–274.
Milner, S. T. see McLeish, T. C. B.: Vol. 143, pp. 195–256.
Mison, P. and *Sillion, B.*: Thermosetting Oligomers Containing Maleimides and Nadiimides End-Groups. Vol. 140, pp. 137–180.
Miyasaka, K.: PVA-Iodine Complexes: Formation, Structure and Properties. Vol. 108. pp. 91–130.
Miller, R. D. see Hedrick, J. L.: Vol. 141, pp. 1–44.
Minko, S. see Rühe, J.: Vol. 165, pp. 79–150.
Möhwald, H., Menzel, H., Helm, C. A. and *Stamm, M.*: Lipid and Polyampholyte Monolayers to Study Polyelectrolyte Interactions and Structure at Interfaces. Vol. 165, pp. 151–175.
Monkenbusch, M. see Richter, D.: Vol. 174, in press
Monnerie, L. see Bahar, I.: Vol. 116, pp. 145–206.
Mori, H. see Bohrisch, J.: Vol. 165, pp. 1–41.
Morishima, Y.: Photoinduced Electron Transfer in Amphiphilic Polyelectrolyte Systems. Vol. 104, pp. 51–96.
Morton M. see Quirk, R. P: Vol. 153, pp. 67–162.
Motornov, M. see Rühe, J.: Vol. 165, pp. 79–150.
Mours, M. see Winter, H. H.: Vol. 134, pp. 165–234.
Müllen, K. see Scherf, U.: Vol. 123, pp. 1–40.
Müller, A. H. E. see Bohrisch, J.: Vol. 165, pp. 1–41.
Müller, A. H. E. see Förster, S.: Vol. 166, pp. 173–210.
Müller, M. see Thünemann, A. F.: Vol. 166, pp. 113–171.
Müller-Plathe, F. see Gusev, A. A.: Vol. 116, pp. 207–248.
Müller-Plathe, F. see Baschnagel, J.: Vol. 152, p. 41–156.
Mukerherjee, A. see Biswas, M.: Vol. 115, pp. 89–124.
Munz, M., Cappella, B., Sturm, H., Geuss, M. and *Schulz, E.*: Materials Contrasts and Nanolithography Techniques in Scanning Force Microscopy (SFM) and their Application to Polymers and Polymer Composites. Vol. 164, pp. 87–210.

Murat, M. see Baschnagel, J.: Vol. 152, p. 41–156.
Mylnikov, V.: Photoconducting Polymers. Vol. 115, pp. 1–88.

Nagy, A. see Majoros, I.: Vol. 112, pp. 1–11.
Naka, K. see Uemura, T.: Vol. 167, pp. 81–106.
Nakamura, A. see Mashima, K.: Vol. 133, pp. 1 52.
Nukayama, Y. see Mashima, K.: Vol. 133, pp. 1–52.
Narasinham, B. and *Peppas, N. A.*: The Physics of Polymer Dissolution: Modeling Approaches and Experimental Behavior. Vol. 128, pp. 157–208.
Nechaev, S. see Grosberg, A.: Vol. 106, pp. 1–30.
Neoh, K. G. see Kang, E. T.: Vol. 106, pp. 135–190.
Netz, R.R. see Holm, C.: Vol. 166, pp. 67–111.
Netz, R.R. see Rühe, J.: Vol. 165, pp. 79–150.
Newman, S. M. see Anseth, K. S.: Vol. 122, pp. 177–218.
Nijenhuis, K. te: Thermoreversible Networks. Vol. 130, pp. 1–252.
Ninan, K. N. see Reghunadhan Nair, C.P.: Vol. 155, pp. 1–99.
Nishi, T. see Jinnai, H.: Vol. 170, pp. 115–167.
Nishikawa, Y. see Jinnai, H.: Vol. 170, pp. 115–167.
Noid, D. W. see Otaigbe, J. U.: Vol. 154, pp. 1–86.
Noid, D. W. see Sumpter, B. G.: Vol. 116, pp. 27–72.
Nomura, M., Tobita, H. and *Suzuki, K.*: Emulsion Polymerization: Kinetic and Mechanistic Aspects. Vol. 175, pp. 1–128.
Novac, B. see Grubbs, R.: Vol. 102, pp. 47–72.
Novikov, V. V. see Privalko, V. P.: Vol. 119, pp. 31–78.
Nummila-Pakarinen, A. see Knuuttila, H.: Vol. 169, pp. 13–27.

O'Brien, D. F., Armitage, B. A., Bennett, D. E. and *Lamparski, H. G.*: Polymerization and Domain Formation in Lipid Assemblies. Vol. 126, pp. 53–84.
Ogasawara, M.: Application of Pulse Radiolysis to the Study of Polymers and Polymerizations. Vol.105, pp. 37–80.
Okabe, H. see Matsushige, K.: Vol. 125, pp. 147–186.
Okada, M.: Ring-Opening Polymerization of Bicyclic and Spiro Compounds. Reactivities and Polymerization Mechanisms. Vol. 102, pp. 1–46.
Okano, T.: Molecular Design of Temperature-Responsive Polymers as Intelligent Materials. Vol. 110, pp. 179–198.
Okay, O. see Funke, W.: Vol. 136, pp. 137–232.
Onuki, A.: Theory of Phase Transition in Polymer Gels. Vol. 109, pp. 63–120.
Oppermann, W. see Holm, C.: Vol. 166, pp. 1–27.
Oppermann, W. see Volk, N.: Vol. 166, pp. 29–65.
Osad'ko, I. S.: Selective Spectroscopy of Chromophore Doped Polymers and Glasses. Vol. 114, pp. 123–186.
Osukuda, K. and *Takeuchi, D.*: Coordination Polymerization of Dienes, Allenes, and Methylenecycloalkanes. Vol. 171, pp. 137–194.
Otaigbe, J. U., Barnes, M. D., Fukui, K., Sumpter, B. G. and *Noid, D. W.*: Generation, Characterization, and Modeling of Polymer Micro- and Nano-Particles. Vol. 154, pp. 1–86.
Otsu, T. and *Matsumoto, A.*: Controlled Synthesis of Polymers Using the Iniferter Technique: Developments in Living Radical Polymerization. Vol. 136, pp. 75–138.

de Pablo, J. J. see Leontidis, E.: Vol. 116, pp. 283–318.
Padias, A. B. see Penelle, J.: Vol. 102, pp. 73–104.

Pascault, J.-P. see *Williams, R. J. J.*: Vol. 128, pp. 95–156.
Pasch, H.: Analysis of Complex Polymers by Interaction Chromatography. Vol. 128, pp. 1–46.
Pasch, H.: Hyphenated Techniques in Liquid Chromatography of Polymers. Vol. 150, pp. 1–66.
Paul, W. see *Baschnagel, J.*: Vol. 152, p. 41–156.
Penczek, P. see *Batog, A. E.*: Vol. 144, pp. 49–114.
Penczek, P. see *Bogdal, D.*: Vol. 163, pp. 193–263.
Penelle, J., Hall, H. K., Padias, A. B. and *Tanaka, H.*: Captodative Olefins in Polymer Chemistry. Vol. 102, pp. 73–104.
Peppas, N. A. see *Bell, C. L.*: Vol. 122, pp. 125–176.
Peppas, N. A. see *Hassan, C. M.*: Vol. 153, pp. 37–65.
Peppas, N. A. see *Narasimhan, B.*: Vol. 128, pp. 157–208.
Pet'ko, I. P. see *Batog, A. E.*: Vol. 144, pp. 49–114.
Pheyghambarian, N. see *Kippelen, B.*: Vol. 161, pp. 87–156.
Pichot, C. see *Hunkeler, D.*: Vol. 112, pp. 115–134.
Pielichowski, J. see *Bogdal, D.*: Vol. 163, pp. 193–263.
Pieper, T. see *Kilian, H. G.*: Vol. 108, pp. 49–90.
Pispas, S. see *Pitsikalis, M.*: Vol. 135, pp. 1–138.
Pispas, S. see *Hadjichristidis, N.*: Vol. 142, pp. 71–128.
Pitsikalis, M., Pispas, S., Mays, J. W. and *Hadjichristidis, N.*: Nonlinear Block Copolymer Architectures. Vol. 135, pp. 1–138.
Pitsikalis, M. see *Hadjichristidis, N.*: Vol. 142, pp. 71–128.
Pleul, D. see *Spange, S.*: Vol. 165, pp. 43–78.
Plummer, C. J. G.: Microdeformation and Fracture in Bulk Polyolefins. Vol. 169, pp. 75–119.
Pötschke, D. see *Dingenouts, N.*: Vol 144, pp. 1–48.
Pokrovskii, V. N.: The Mesoscopic Theory of the Slow Relaxation of Linear Macromolecules. Vol. 154, pp. 143–219.
Pospíšil, J.: Functionalized Oligomers and Polymers as Stabilizers for Conventional Polymers. Vol. 101, pp. 65–168.
Pospíšil, J.: Aromatic and Heterocyclic Amines in Polymer Stabilization. Vol. 124, pp. 87–190.
Powers, A. C. see *Prokop, A.*: Vol. 136, pp. 53–74.
Prasad, P. N. see *Lin, T.-C.*: Vol. 161, pp. 157–193.
Priddy, D. B.: Recent Advances in Styrene Polymerization. Vol. 111, pp. 67–114.
Priddy, D. B.: Thermal Discoloration Chemistry of Styrene-co-Acrylonitrile. Vol. 121, pp. 123–154.
Privalko, V. P. and *Novikov, V. V.*: Model Treatments of the Heat Conductivity of Heterogeneous Polymers. Vol. 119, pp. 31–78.
Prociak, A. see *Bogdal, D.*: Vol. 163, pp. 193–263.
Prokop, A., Hunkeler, D., DiMari, S., Haralson, M. A. and *Wang, T. G.*: Water Soluble Polymers for Immunoisolation I: Complex Coacervation and Cytotoxicity. Vol. 136, pp. 1–52.
Prokop, A., Hunkeler, D., Powers, A. C., Whitesell, R. R. and *Wang, T. G.*: Water Soluble Polymers for Immunoisolation II: Evaluation of Multicomponent Microencapsulation Systems. Vol. 136, pp. 53–74.
Prokop, A., Kozlov, E., Carlesso, G. and *Davidsen, J. M.*: Hydrogel-Based Colloidal Polymeric System for Protein and Drug Delivery: Physical and Chemical Characterization, Permeability Control and Applications. Vol. 160, pp. 119–174.
Pruitt, L. A.: The Effects of Radiation on the Structural and Mechanical Properties of Medical Polymers. Vol. 162, pp. 65–95.
Pudavar, H. E. see *Lin, T.-C.*: Vol. 161, pp. 157–193.

Pukánszky, B. and *Fekete, E.*: Adhesion and Surface Modification. Vol. 139, pp. 109–154.
Putnam, D. and *Kopecek, J.*: Polymer Conjugates with Anticancer Acitivity. Vol. 122, pp. 55–124.

Quirk, R. P., Yoo, T., Lee, Y., M., Kim, J. and *Lee, B.*: Applications of 1,1-Diphenylethylene Chemistry in Anionic Synthesis of Polymers with Controlled Structures. Vol. 153, pp. 67–162.

Ramaraj, R. and *Kaneko, M.*: Metal Complex in Polymer Membrane as a Model for Photosynthetic Oxygen Evolving Center. Vol. 123, pp. 215–242.
Rangarajan, B. see Scranton, A. B.: Vol. 122, pp. 1–54.
Ranucci, E. see Söderqvist Lindblad, M.: Vol. 157, pp. 139–161.
Raphaël, E. see Léger, L.: Vol. 138, pp. 185–226.
Reddinger, J. L. and *Reynolds, J. R.*: Molecular Engineering of p-Conjugated Polymers. Vol. 145, pp. 57–122.
Reghunadhan Nair, C. P., Mathew, D. and *Ninan, K. N.*: Cyanate Ester Resins, Recent Developments. Vol. 155, pp. 1–99.
Reichert, K. H. see Hunkeler, D.: Vol. 112, pp. 115–134.
Rehahn, M., Mattice, W. L. and *Suter, U. W.*: Rotational Isomeric State Models in Macromolecular Systems. Vol. 131/132, pp. 1–475.
Rehahn, M. see Bohrisch, J.: Vol. 165, pp. 1–41.
Rehahn, M. see Holm, C.: Vol. 166, pp. 1–27.
Reineker, P. see Holm, C.: Vol. 166, pp. 67–111.
Reitberger, T. see Jacobson, K.: Vol. 169, pp. 151–176.
Reynolds, J. R. see Reddinger, J. L.: Vol. 145, pp. 57–122.
Richter, D. see Ewen, B.: Vol. 134, pp.1–130.
Richter, D., Monkenbusch, M. and *Colmenero J.*: Neutron Spin Echo in Polymer Systems. Vol. 174, in press
Riegler, S. see Trimmel, G.: Vol. 176, pp. 43–87.
Risse, W. see Grubbs, R.: Vol. 102, pp. 47–72.
Rivas, B. L. and *Geckeler, K. E.*: Synthesis and Metal Complexation of Poly(ethyleneimine) and Derivatives.Vol. 102, pp. 171–188.
Roberts, G. W. see Kennedy, K. A.: Vol. 175, pp. 329–346.
Robin, J. J.: The Use of Ozone in the Synthesis of New Polymers and the Modification of Polymers. Vol. 167, pp. 35–79.
Robin, J. J. see Boutevin, B.: Vol. 102, pp. 105–132.
Roe, R.-J.: MD Simulation Study of Glass Transition and Short Time Dynamics in Polymer Liquids. Vol. 116, pp. 111–114.
Roovers, J. and *Comanita, B.*: Dendrimers and Dendrimer-Polymer Hybrids. Vol. 142, pp. 179–228.
Rothon, R. N.: Mineral Fillers in Thermoplastics: Filler Manufacture and Characterisation. Vol. 139, pp. 67–108.
Rozenberg, B. A. see Williams, R. J. J.: Vol. 128, pp. 95–156.
Rühe, J., Ballauff, M., Biesalski, M., Dziezok, P., Gröhn, F., Johannsmann, D., Houbenov, N., Hugenberg, N., Konradi, R., Minko, S., Motornov, M., Netz, R. R., Schmidt, M., Seidel, C., Stamm, M., Stephan, T., Usov, D. and *Zhang, H.*: Polyelectrolyte Brushes. Vol. 165, pp. 79–150.
Ruckenstein, E.: Concentrated Emulsion Polymerization. Vol. 127, pp. 1–58.
Ruiz-Taylor, L. see Mathieu, H. J.: Vol. 162, pp. 1–35.
Rusanov, A. L.: Novel Bis (Naphtalic Anhydrides) and Their Polyheteroarylenes with Improved Processability. Vol. 111, pp. 115–176.

Russel, T. P. see Hedrick, J. L.: Vol. 141, pp. 1–44.
Russum, J. P. see Schork, F. J.: Vol. 175, pp. 129–255.
Rychly, J. see Lazár, M.: Vol. 102, pp. 189–222.
Ryner, M. see Stridsberg, K. M.: Vol. 157, pp. 27–51.
Ryzhov, V. A. see Bershtein, V. A.: Vol. 114, pp. 43–122.

Sabsai, O. Y. see Barshtein, G. R.: Vol. 101, pp. 1–28.
Saburov, V. V. see Zubov, V. P.: Vol. 104, pp. 135–176.
Saito, S., Konno, M. and *Inomata, H.*: Volume Phase Transition of N-Alkylacrylamide Gels. Vol. 109, pp. 207–232.
Samsonov, G. V. and *Kuznetsova, N. P.*: Crosslinked Polyelectrolytes in Biology. Vol. 104, pp. 1–50.
Santa Cruz, C. see Baltá-Calleja, F. J.: Vol. 108, pp. 1–48.
Santos, S. see Baschnagel, J.: Vol. 152, p. 41–156.
Sato, T. and *Teramoto, A.*: Concentrated Solutions of Liquid-Christalline Polymers. Vol. 126, pp. 85–162.
Schaller, C. see Bohrisch, J.: Vol. 165, pp. 1–41.
Schäfer R. see Köhler, W.: Vol. 151, pp. 1–59.
Scherf, U. and *Müllen, K.*: The Synthesis of Ladder Polymers. Vol. 123, pp. 1–40.
Schmidt, M. see Förster, S.: Vol. 120, pp. 51–134.
Schmidt, M. see Rühe, J.: Vol. 165, pp. 79–150.
Schmidt, M. see Volk, N.: Vol. 166, pp. 29–65.
Scholz, M.: Effects of Ion Radiation on Cells and Tissues. Vol. 162, pp. 97–158.
Schopf, G. and *Koßmehl, G.*: Polythiophenes – Electrically Conductive Polymers. Vol. 129, pp. 1–145.
Schork, F. J., Luo, Y., Smulders, W., Russum, J. P., Butté, A. and *Fontenot, K.*: Miniemulsion Polymerization. Vol. 175, pp. 127–255.
Schulz, E. see Munz, M.: Vol. 164, pp. 97–210.
Seppälä, J. see Löfgren, B.: Vol. 169, pp. 1–12.
Sturm, H. see Munz, M.: Vol. 164, pp. 87–210.
Schweizer, K. S.: Prism Theory of the Structure, Thermodynamics, and Phase Transitions of Polymer Liquids and Alloys. Vol. 116, pp. 319–378.
Scranton, A. B., Rangarajan, B. and *Klier, J.*: Biomedical Applications of Polyelectrolytes. Vol. 122, pp. 1–54.
Sefton, M. V. and *Stevenson, W. T. K.*: Microencapsulation of Live Animal Cells Using Polycrylates. Vol.107, pp. 143–198.
Seidel, C. see Holm, C.: Vol. 166, pp. 67–111.
Seidel, C. see Rühe, J.: Vol. 165, pp. 79–150.
Shamanin, V. V.: Bases of the Axiomatic Theory of Addition Polymerization. Vol. 112, pp. 135–180.
Sheiko, S. S.: Imaging of Polymers Using Scanning Force Microscopy: From Superstructures to Individual Molecules. Vol. 151, pp. 61–174.
Sherrington, D. C. see Cameron, N. R.,Vol. 126, pp. 163–214.
Sherrington, D. C. see Lin, J.: Vol. 111, pp. 177–220.
Sherrington, D. C. see Steinke, J.: Vol. 123, pp. 81–126.
Shibayama, M. see Tanaka, T.: Vol. 109, pp. 1–62.
Shiga, T.: Deformation and Viscoelastic Behavior of Polymer Gels in Electric Fields. Vol. 134, pp. 131–164.
Shim, H.-K. and *Jin, J.*: Light-Emitting Characteristics of Conjugated Polymers. Vol. 158, pp. 191–241.
Shoda, S. see Kobayashi, S.: Vol. 121, pp. 1–30.

Siegel, R. A.: Hydrophobic Weak Polyelectrolyte Gels: Studies of Swelling Equilibria and Kinetics. Vol. 109, pp. 233–268.
Silvestre, F. see Calmon-Decriaud, A.: Vol. 207, pp. 207–226.
Sillion, B. see Mison, P.: Vol. 140, pp. 137–180.
Simon, F. see Spange, S.: Vol. 165, pp. 43–78.
Singh, R. P. see Sivaram, S.: Vol. 101, pp. 169–216.
Singh, R. P. see Desai, S. M.: Vol. 169, pp. 231–293.
Sinha Ray, S. see Biswas, M: Vol. 155, pp. 167–221.
Sivaram, S. and *Singh, R. P.*: Degradation and Stabilization of Ethylene-Propylene Copolymers and Their Blends: A Critical Review. Vol. 101, pp. 169–216.
Slugovc, C. see Trimmel, G.: Vol. 176, pp. 43–87.
Smulders, W. see Schork, F. J.: Vol. 175, pp. 129–255.
Söderqvist Lindblad, M., Liu, Y., Albertsson, A.-C., Ranucci, E. and *Karlsson, S.*: Polymer from Renewable Resources.Vol. 157, pp. 139–161.
Spange, S., Meyer, T., Voigt, I., Eschner, M., Estel, K., Pleul, D. and *Simon, F.*: Poly(Vinylformamide-co-Vinylamine)/Inorganic Oxid Hybrid Materials. Vol. 165, pp. 43–78.
Stamm, M. see Möhwald, H.: Vol. 165, pp. 151–175.
Stamm, M. see Rühe, J.: Vol. 165, pp. 79–150.
Starodybtzev, S. see Khokhlov, A.: Vol. 109, pp. 121–172.
Stegeman, G. I. see Canva, M.: Vol. 158, pp. 87–121.
Steinke, J., Sherrington, D. C. and *Dunkin, I. R.*: Imprinting of Synthetic Polymers Using Molecular Templates. Vol. 123, pp. 81–126.
Stelzer, F. see Trimmel, G.: Vol. 176, pp. 43–87.
Stenberg, B. see Jacobson, K.: Vol. 169, pp. 151–176.
Stenzenberger, H. D.: Addition Polyimides. Vol. 117, pp. 165–220.
Stephan, T. see Rühe, J.: Vol. 165, pp. 79–150.
Stevenson,W. T. K. see Sefton, M. V.: Vol. 107, pp. 143–198.
Stridsberg, K. M., Ryner, M. and *Albertsson, A.-C.*: Controlled Ring-Opening Polymerization: Polymers with Designed Macromoleculars Architecture. Vol. 157, pp. 27–51.
Sturm, H. see Munz, M.: Vol. 164, pp. 87–210.
Suematsu, K.: Recent Progress of Gel Theory: Ring, Excluded Volume, and Dimension. Vol. 156, pp. 136–214.
Sugimoto, H. and *Inoue, S.*: Polymerization by Metalloporphyrin and Related Complexes. Vol. 146, pp. 39–120.
Suginome, M. and *Ito, Y.*: Transition Metal-Mediated Polymerization of Isocyanides. Vol. 171, pp. 77–136.
Sumpter, B. G., Noid, D. W., Liang, G. L. and *Wunderlich, B.*: Atomistic Dynamics of Macromolecular Crystals. Vol. 116, pp. 27–72.
Sumpter, B. G. see Otaigbe, J. U.: Vol. 154, pp. 1–86.
Sun, H.-B. and *Kawata, S.*: Two-Photon Photopolymerization and 3D Lithographic Microfabrication. Vol. 170, pp. 169–273.
Suter, U. W. see Gusev, A. A.: Vol. 116, pp. 207–248.
Suter, U. W. see Leontidis, E.: Vol. 116, pp. 283–318.
Suter, U. W. see Rehahn, M.: Vol. 131/132, pp. 1–475.
Suter, U. W. see Baschnagel, J.: Vol. 152, p. 41–156.
Suzuki, A.: Phase Transition in Gels of Sub-Millimeter Size Induced by Interaction with Stimuli. Vol. 110, pp. 199–240.
Suzuki, A. and *Hirasa, O.*: An Approach to Artifical Muscle by Polymer Gels due to Micro-Phase Separation. Vol. 110, pp. 241–262.
Suzuki, K. see Nomura, M.: Vol. 175, pp. 1–128.
Swiatkiewicz, J. see Lin, T.-C.: Vol. 161, pp. 157–193.

Tagawa, S.: Radiation Effects on Ion Beams on Polymers. Vol. 105, pp. 99–116.
Takata, T., Kihara, N. and *Furusho, Y.:* Polyrotaxanes and Polycatenanes: Recent Advances in Syntheses and Applications of Polymers Comprising of Interlocked Structures. Vol. 171, pp. 1–75.
Takeuchi, D. see Osakada, K.: Vol. 171, pp. 137–194.
Tan, K. L. see Kang, E. T.: Vol. 106, pp. 135–190.
Tanaka, H. and *Shibayama, M.:* Phase Transition and Related Phenomena of Polymer Gels. Vol. 109, pp. 1–62.
Tanaka, T. see Penelle, J.: Vol. 102, pp. 73–104.
Tauer, K. see Guyot, A.: Vol. 111, pp. 43–66.
Teramoto, A. see Sato, T.: Vol. 126, pp. 85–162.
Terent'eva, J. P. and *Fridman, M. L.:* Compositions Based on Aminoresins. Vol. 101, pp. 29–64.
Theodorou, D. N. see Dodd, L. R.: Vol. 116, pp. 249–282.
Thomson, R. C., Wake, M. C., Yaszemski, M. J. and *Mikos, A. G.:* Biodegradable Polymer Scaffolds to Regenerate Organs. Vol. 122, pp. 245–274.
Thünemann, A. F., Müller, M., Dautzenberg, H., Joanny, J.-F. and *Löwen, H.:* Polyelectrolyte complexes. Vol. 166, pp. 113–171.
Tieke, B. see v. Klitzing, R.: Vol. 165, pp. 177–210.
Tobita, H. see Nomura, M.: Vol. 175, pp. 1–128.
Tokita, M.: Friction Between Polymer Networks of Gels and Solvent. Vol. 110, pp. 27–48.
Traser, S. see Bohrisch, J.: Vol. 165, pp. 1–41.
Tries, V. see Baschnagel, J.: Vol. 152, p. 41–156.
Trimmel, G., Riegler, S., Fuchs, G., Slugovc, C. and *Stelzer, F.:* Liquid Crystalline Polymers by Metathesis Polymerization. Vol. 176, pp. 43–87.
Tsuruta, T.: Contemporary Topics in Polymeric Materials for Biomedical Applications. Vol. 126, pp. 1–52.

Uemura, T., Naka, K. and *Chujo, Y.:* Functional Macromolecules with Electron-Donating Dithiafulvene Unit. Vol. 167, pp. 81–106.
Usov, D. see Rühe, J.: Vol. 165, pp. 79–150.
Uyama, H. see Kobayashi, S.: Vol. 121, pp. 1–30.
Uyama, Y: Surface Modification of Polymers by Grafting. Vol. 137, pp. 1–40.

Varma, I. K. see Albertsson, A.-C.: Vol. 157, pp. 99–138.
Vasilevskaya, V. see Khokhlov, A.: Vol. 109, pp. 121–172.
Vaskova, V. see Hunkeler, D.: Vol.: 112, pp. 115–134.
Verdugo, P.: Polymer Gel Phase Transition in Condensation-Decondensation of Secretory Products. Vol. 110, pp. 145–156.
Vettegren, V. I. see Bronnikov, S. V.: Vol. 125, pp. 103–146.
Vilgis, T. A. see Holm, C.: Vol. 166, pp. 67–111.
Viovy, J.-L. and *Lesec, J.:* Separation of Macromolecules in Gels: Permeation Chromatography and Electrophoresis. Vol. 114, pp. 1–42.
Vlahos, C. see Hadjichristidis, N.: Vol. 142, pp. 71–128.
Voigt, I. see Spange, S.: Vol. 165, pp. 43–78.
Volk, N., Vollmer, D., Schmidt, M., Oppermann, W. and *Huber, K.:* Conformation and Phase Diagrams of Flexible Polyelectrolytes. Vol. 166, pp. 29–65.
Volksen, W.: Condensation Polyimides: Synthesis, Solution Behavior, and Imidization Characteristics. Vol. 117, pp. 111–164.
Volksen, W. see Hedrick, J. L.: Vol. 141, pp. 1–44.
Volksen, W. see Hedrick, J. L.: Vol. 147, pp. 61–112.
Vollmer, D. see Volk, N.: Vol. 166, pp. 29–65.

Wagener, K. B. see Baughman, T. W.: Vol 176, pp. 1–42.
Wake, M. C. see Thomson, R. C.: Vol. 122, pp. 245–274.
Wandrey C., Hernández-Barajas, J. and *Hunkeler, D.*: Diallyldimethylammonium Chloride and its Polymers. Vol. 145, pp. 123–182.
Wang, K. L. see Cussler, E. L.: Vol. 110, pp. 67–80.
Wang, S.-Q.: Molecular Transitions and Dynamics at Polymer/Wall Interfaces: Origins of Flow Instabilities and Wall Slip. Vol. 138, pp. 227–276.
Wang, S.-Q. see Bhargava, R.: Vol. 163, pp. 137–191.
Wang, T. G. see Prokop, A.: Vol. 136, pp. 1–52; 53–74.
Wang, X. see Lin, T.-C.: Vol. 161, pp. 157–193.
Webster, O. W.: Group Transfer Polymerization: Mechanism and Comparison with Other Methods of Controlled Polymerization of Acrylic Monomers. Vol. 167, pp. 1–34.
Whitesell, R. R. see Prokop, A.: Vol. 136, pp. 53–74.
Williams, R. J. J., Rozenberg, B. A. and *Pascault, J.-P.*: Reaction Induced Phase Separation in Modified Thermosetting Polymers. Vol. 128, pp. 95–156.
Winkler, R. G. see Holm, C.: Vol. 166, pp. 67–111.
Winter, H. H. and *Mours, M.*: Rheology of Polymers Near Liquid-Solid Transitions. Vol. 134, pp. 165–234.
Wittmeyer, P. see Bohrisch, J.: Vol. 165, pp. 1–41.
Wu, C.: Laser Light Scattering Characterization of Special Intractable Macromolecules in Solution. Vol 137, pp. 103–134.
Wunderlich, B. see Sumpter, B. G.: Vol. 116, pp. 27–72.

Xiang, M. see Jiang, M.: Vol. 146, pp. 121–194.
Xie, T. Y. see Hunkeler, D.: Vol. 112, pp. 115–134.
Xu, Z., Hadjichristidis, N., Fetters, L. J. and *Mays, J. W.*: Structure/Chain-Flexibility Relationships of Polymers. Vol. 120, pp. 1–50.

Yagci, Y. and *Endo, T.*: N-Benzyl and N-Alkoxy Pyridium Salts as Thermal and Photochemical Initiators for Cationic Polymerization. Vol. 127, pp. 59–86.
Yannas, I. V.: Tissue Regeneration Templates Based on Collagen-Glycosaminoglycan Copolymers. Vol. 122, pp. 219–244.
Yang, J. S. see Jo, W. H.: Vol. 156, pp. 1–52.
Yamaoka, H.: Polymer Materials for Fusion Reactors. Vol. 105, pp. 117–144.
Yasuda, H. and *Ihara, E.*: Rare Earth Metal-Initiated Living Polymerizations of Polar and Nonpolar Monomers. Vol. 133, pp. 53–102.
Yaszemski, M. J. see Thomson, R. C.: Vol. 122, pp. 245–274.
Yoo, T. see Quirk, R. P.: Vol. 153, pp. 67–162.
Yoon, D. Y. see Hedrick, J. L.: Vol. 141, pp. 1–44.
Yoshida, H. and *Ichikawa, T.*: Electron Spin Studies of Free Radicals in Irradiated Polymers. Vol. 105, pp. 3–36.

Zhang, H. see Rühe, J.: Vol. 165, pp. 79–150.
Zhang, Y.: Synchrotron Radiation Direct Photo Etching of Polymers. Vol. 168, pp. 291–340.
Zhou, H. see Jiang, M.: Vol. 146, pp. 121–194.
Zubov, V . P., Ivanov, A. E. and *Saburov, V. V.*: Polymer-Coated Adsorbents for the Separation of Biopolymers and Particles. Vol. 104, pp. 135–176.

Subject Index

Actuator 76
Acyclic diene metathesis (ADMET) 1–7, 44, 81, 90
α-Addition 97, 103
β-Addition 97, 104
ADMET 1–7, 44, 81, 90
–, copolymerizations 7
–, depolymerization 6
–, polymerization 90
Alignment, mechanical 58
1-Alkyne polymerization 90, 93, 115
Alternating diene metathesis polycondensation (ALTMET) 23, 44, 84
Amphiphilic structure 114
Anchoring group 48
Anionic polymerization 101
Anisotropic alignment 48
Atom transfer radical polymerization (ATRP) 15, 17, 75
Azobenzene 64

Biphenyl-based mesogens 48
Birefringence 56
Block copolymers 65, 75

Carbosilane polymers 32
Chauvin mechanism 4
Chiral polymers 27
Chlorinated polyethylene (CPE) 13
Cholesteryl group 67
Class VI living polymerization system 111
CNO-face 104
Conducting polymer 91
Conjugated polymers 25
Cross metathesis (CM) 3, 26
Cyanostilbenyl groups 67
Cyclopolymerization 100, 108, 110

Diads 54
Diene metathesis polycondensation, alternating (ALTMET) 84
Diethyl dipropargylmalonate (DEDPM) 90, 101, 105, 108, 110, 114, 115
1,2-Diethynylbenzene 101
Differential scanning calorimetry (DSC) 10, 12, 14, 21
Durham route 95

Effective conjugation length 90, 108
Element of symmetry 107
Enyne metathesis 109
Ethynylferrocene 92–96, 99
Ethynylruthenocene 93, 96

^{57}Fe-Mössbauer spectroscopy 99
Feast monomer 95
4-Ferrocenylethynyl-4′ethynyltolan 95
Fluorinated side chains 65, 71
Free radical copolymerization 6, 10

Graft copolymers 15
Grubbs-Herrmann catalyst 109
Grubbs-Hoveyda catalyst 109

1,6-Heptadiyne 89, 100, 101, 108, 110, 115
Hoveyda catalyst 110

Initiators 46, 54
–, efficiency 108
Inositol 30
Isomer shift (IS) 90

Latent reactive polymers 35
LC elastomers 58
LCP 43
–, lyotropic 44

Subject Index

LCP
–, main chain 81
Liquid crystals 43

Macroinitiated polymerization 18
Macroinitiator 21
Macromolecular substitution 34
Magnetic orientation 56
MALDI-TOF spectroscopy 109
MCLCPs 45
Mesogens 44
–, 2,5-bis[(4'-n-alkoxybenzoyl)oxy]phenyl 68
–, [(4'-cyanobiphenyl-4-yl-)oxy] 49
–, biphenyl-based 48
–, calamitic/discotic 45
–, density 48, 52
–, discotic 45, 77
–, laterally-attached 67
Metathesis, LCPs 43
Metathesis reaction 3
Metathesis telomerization 26
Micelles 45, 112
Microdomains 61
Microphase separation 65
Molybdenum 46, 54, 90, 96, 109

Network 36
p-Nitrostilbene 63

Odd-even effect 50, 52, 69
Olefin metathesis 3

Phase separation 62
Phosphazenes 23
Photo-switchable 64
Poly(cyclohex-1-ene-methylidene) 102, 103, 108
Poly(cyclopent-1-enyle-1-vinylene) 102–108
Poly(DEDPM) 104, 114, 115
Poly(dimethyl siloxane) 33
Poly(ethynylferrocene) 94
Poly(ethynylruthenocene) 94
Poly(2-oxazoline) 112
Poly(p-phenylene) 92
Poly(p-phenylene vinylene) (PPV) 25, 90
Poly(pyrrole) 92
Poly(thiazole) 92

Poly(thiophene) 92
Polyethylene 1, 6
–, methyl branched 7
–, modeling 8
Polyoctenamer 21, 33

Quinuclidine 105, 108

Rate constant for initiation (k_i)/polymerization (k_p) 93
Relaxation 54
Rheological behavior 54
Rhodium 90
Ring-closing metathesis (RCM) 3, 109
Ring-opening cross metathesis 109
Ring-opening insertion metathesis polymerization (ROIMP) 22
Ring-opening metathesis (ROM) 3
Ring-opening metathesis polymerization (ROMP) 3, 5, 43, 45, 90, 109
Ring-opening polymerization (ROP) 15
anti-/syn-Rotamer 104, 105
Ruthenium 46

Schrock initiator 90, 93, 109
SCLCPs 43, 47
– block copolymers 61
Side chains, dendritic 78
– –, fluorinated 65, 71
Side-chain crystallization 50
Siloxane-terminated monomers 74
Sodium dodecylsolfonate 112
Spacer length, influence 49, 52
Spacers 47, 48
Supramolecular arrangement 78

Tacticity 54
Telechelic polymers 18, 33
Thermal history 53
Triblock copolymers 18, 21
o-Trimethylsilylphenylacetylene (o-TMSPA) 90, 96
Triphenylenes 77
Tungsten 90

Uniaxial orientation 56

Z/E ratio 54
Ziegler-Natta catalysts 101

Printing: Krips bv, Meppel
Binding: Litges & Dopf, Heppenheim